T0233305

Cambridge Elements ≡

Elements in the Philosophy of Biology
edited by
Grant Ramsey
KU Leuven
Michael Ruse
Florida State University

STEM CELLS

Melinda Bonnie Fagan
University of Utah

CAMBRIDGE
UNIVERSITY PRESS

CAMBRIDGE
UNIVERSITY PRESS

University Printing House, Cambridge CB2 8BS, United Kingdom

One Liberty Plaza, 20th Floor, New York, NY 10006, USA

477 Williamstown Road, Port Melbourne, VIC 3207, Australia

314–321, 3rd Floor, Plot 3, Splendor Forum, Jasola District Centre,
New Delhi – 110025, India

79 Anson Road, #06–04/06, Singapore 079906

Cambridge University Press is part of the University of Cambridge.

It furthers the University's mission by disseminating knowledge in the pursuit of
education, learning, and research at the highest international levels of excellence.

www.cambridge.org
Information on this title: www.cambridge.org/9781108741712
DOI: 10.1017/9781108680783

© Melinda Bonnie Fagan 2021

This publication is in copyright. Subject to statutory exception
and to the provisions of relevant collective licensing agreements,
no reproduction of any part may take place without the written
permission of Cambridge University Press.

First published 2021

A catalogue record for this publication is available from the British Library.

ISBN 978-1-108-74171-2 Paperback
ISSN 2515-1126 (online)
ISSN 2515-1118 (print)

Cambridge University Press has no responsibility for the persistence or accuracy of
URLs for external or third-party internet websites referred to in this publication
and does not guarantee that any content on such websites is, or will remain,
accurate or appropriate.

Stem Cells

Elements in the Philosophy of Biology

DOI: 10.1017/9781108680783
First published online: May 2021

Melinda Bonnie Fagan
University of Utah

Abstract: What is a stem cell? The question is deceptively simple; the answer seemingly obvious. A stem cell is a *cell* that serves as a *stem*, or point of origin, for something else. But beneath the surface of this simple conjunction is a swirling mass of further questions. What does the stem lead to? What is a stem cell a stem *of*? What connects the stem to its sequel? Is the connection "pre-formed" in the stem cell itself or governed by external factors? The very idea of a stem cell leads directly to fundamental questions about the nature of biological development. Stem cell science, by the same token, raises issues central for the philosophy of science.

Keywords: stem cells, experiment, models, development, regenerative medicine

© Melinda Bonnie Fagan 2021

ISBNs: 9781108741712 (PB), 9781108680783 (OC)
ISSNs: 2515-1126 (online), 2515-1118 (print)

Contents

1 Background   1

2 Stem Cells: The Very Idea   10

3 Finding Stem Cells   35

4 Using Stem Cells   62

  Epilogue   74

  References   76

1 Background

What Is a Stem Cell?

The term itself seems to answer the question: a stem cell is a cell that is also a stem. A *cell* is a basic category of living thing, a fundamental "unit of life." A *stem* is a site of growth – an active source that supports or gives rise to something else. Both concepts are deeply rooted in biological thought, with rich and complex histories. The idea of a stem cell unites them, but the union is neither simple nor straightforward. The answer to this opening question unfolds throughout this and the following sections.

Why should philosophers care about stem cells? There are two main reasons. First, stem cells and related phenomena bear on classic biological questions about the nature of development, regeneration, and individuality. These are central topics for philosophy of biology. Second, the ways scientists gain knowledge about stem cells, and the forms this knowledge takes, are unlike those of theory-oriented sciences. So stem cell biology is a rich, relatively untapped source of insights about central ideas in biological thought and about scientific practices and knowledge. The following sections present some of these insights. As a prelude, a brief introduction to stem cell research is appropriate.

Stem cell phenomena – the things stem cell researchers strive to understand – run the gamut from familiar to outré. Among the former are ongoing processes within our bodies: hair grows, skin is shed and replaced, and our cells gradually turn over. These changes are commonplace yet mysterious in their working. Other species exhibit more dramatic regenerative abilities. Starfish and salamanders can replace severed limbs; worms and plants can regrow an entire body from a fragment. Humans' more modest regenerative powers show in wound healing: bones knit, spilled blood is replaced, torn skin and muscle become whole. Such "self-renewing" processes are usually internal and undetected. But over the past century, innovations in ex vivo cell culture have made visible many interior aspects of our embodied experience.[1] Many striking stem cell phenomena occur in transparent artificial bodies of liquid and glass[2]: embryonic stem cell lines, induced pluripotent stem cells, embryoid bodies, organoids, embryo-like structures, and more. These new experimental products offer unprecedented views of heretofore hidden aspects of organismal development, human and otherwise. Their strangeness and artificiality notwithstanding, novel

---

[1] Landecker (2007) is an excellent cultural history of the practice and significance of culturing cells. Simian and Bissell (2017) is a more narrowly focused review of ex vivo tissue and organ culture.

[2] Or plastic, but that is less poetic.

stem cell phenomena are created so as to better understand (and control) more familiar ones. Primary among these is the invisible process that carries us from birth to death: *biological development*. Stem cells are the generative stuff of our bodies – the developmental matrix of all multicellular organisms. To understand them is to understand how we are formed, heal, and decline with age. To control them is to control those processes. No wonder the promise of stem cells tends to be overhyped! They are keys to the shape of our lives as living organisms. Of course, we do not understand stem cells fully, and our control over them – at least for now – is extremely modest. Yet what we have learned so far is fascinating and rich in philosophical implications.

Understanding stem cells requires looking carefully at stem cell research practices. Those practices, like much of science today, are very technical and specialized. In addition, stem cell biology presents a further challenge in that it does not include theories, explanations, and abstract models of the sort that philosophers traditionally focus on.[3] Stem cell research has no easily identifiable set of core principles, exemplary methods, or results. Instead, the field presents a sprawling diversity of experimental innovations and achievements, characterized in highly technical ways that make their significance hard for outsiders to appreciate. This experimental character offers a bracing challenge to the philosophy of science: How can we approach such a field and glean philosophical insights from it? This Element takes a particular approach to meeting that challenge, reflected in the organization of its sections. The rest of this section introduces the conceptual background of "cell and stem," along with associated ideas of cell type, division, and lineage. This sets the stage for Section 2, which examines and clarifies the idea of a stem cell: the stem cell concept. That clarification, an abstract model, connects directly to central experimental methods and standards of stem cell biology today. Section 3 discusses these experiments, with further insights about forms of knowledge and progress in stem cell research. Section 4 concludes by examining some important uses of stem cells, both realized and hoped for.

Cells, Cell Types, and Differentiation

First, let's explore some conceptual background, sketching the duality of "cell" and "stem" in biological thought. The term *cell* was introduced to science by Robert Hooke in 1665 – but his meaning was not that of biology today. Hooke was referring to the minuscule empty chambers in slices of cork, revealed by his

---

[3] Stem cell research is not unique in its experimental, decentralized, variegated character. Indeed, many areas of life science today are like this, and also relatively neglected by philosophers. Thus some ideas in this book apply beyond stem cells, although my focus here is on that topic.

early forays with the microscope. The study of cells progresses hand-in-hand with advances in microscopy.[4] Hooke, a pioneer of that instrument, intended an analogy with a monk's cell: a small, confined space. The concept of the cell as a living entity came later, codified in nineteenth-century "cell-theory."[5] Theodor Schwann, in *Microscopical Researches into the Accordance in the Structure and Growth of Animals and Plants* (1847), states "the proposition, that there exists one general principle for the formation of all organic productions, and that this principle is the formation of cells," together with conclusions inferred therefrom.[6] One such implication is that cells, the elementary units making up an organism's various tissues and organs, must themselves develop in different ways. That is, organismal development involves (or is based on, or consists in) cells undergoing processes of change. Those processes are represented as different paths or tracks: the pathways of *differentiation*. This implication of early cell theory remains a deep background assumption of stem cell research today.

In following the various pathways of differentiation, cells of an organism become different from one another. Some of these differences can be observed with a simple light microscope. Thus, from the mid-nineteenth century onward, scientists have been confronted by observable variety among the cells making up the body of an animal or plant. The prevailing response to such diversity is to classify cells of multicellular organisms into distinct types according to features they have as individuals, such as size and shape.[7] That is, if cells have similar features in their own right, as individual entities observable under a microscope (or otherwise detectable), then those cells are classified as the same type or kind. Advances in microscopy, genomics, and other methods of cell characterization have led to myriad changes in the suite of features used to classify organismal cells into distinct types. But the basic principle of classification remains the same. Today, different cell types are defined in terms of robust, distinct clusters of molecular, biochemical, morphological, and functional traits. The number of cell types in an organism can range from three to hundreds, depending on the species and on how fine-grained a characterization

---

[4] See Hacking (1983) for a pioneering discussion of the latter. For historical and philosophical studies of cell biology, see Dröscher (2002), Matlin et al. (2018), and Reynolds (2018).

[5] "The cell theory" is credited jointly to Matthias Schlieden and Theodor Schwann. The history of biological thought on cells is more complex than this brief introduction can indicate; see, for example, Reynolds (2007, 2018) for more detailed treatments.

[6] Schwann (1847, 166).

[7] There are actually two distinct classificatory systems for cells: one for single-cell organisms (microbes) and the other for cells making up the bodies of multicellular organisms. For reasons made clear in the next section, this Element deals only with the latter. See O'Malley (2014), and references therein, for more on microbial classification; and Kendig (2016) for more on classification in the life sciences.

is wanted. In principle, any multicellular organism can be decomposed into a heap of its constituent cells, and those can then be clustered into types and subtypes.[8]

Another idea from early cell theory that has settled into the deep conceptual background of biology today is Schwann's principle of formation: that all cells are produced in the same way. This mode of production is *cell division*: every cell is generated from a parent cell through a process of binary division.[9] Working out the stages of cell division and coordinating these with inherited material localized to chromosomes was a towering achievement of late-nineteenth- and early-twentieth-century biology. Together with breakthroughs on the process of fertilization, studies of cell division yielded a general, universal account of the reproduction and development of all organisms, grounded on the cellular processes of gamete formation (meiosis) and binary division (mitosis).[10] These ideas today operate as background assumptions for many biological fields, including stem cell research. Although rarely stated explicitly, they provide clear criteria for the existence and individuation of cells:

- An individual cell's existence begins with a cell division event (as one of a pair of offspring cells).
- An individual cell's existence ends with either a second division event (producing two offspring) or cell death (and no offspring).[11]

The cell membrane, a universal feature of cells, makes these "conditions of existence" even more precise and clear. Every cell is bounded by an enclosing membrane.[12] This makes for a topological criterion of cell individuality; a cell is

---

[8] Cells are not the only constituents of multicellular organisms; nor is cell-level decomposition the only possible or scientifically significant breakdown of these organisms. The idea that a multicellular organism is an assemblage or "society" of cells was prevalent in late-nineteenth-century biology – when key concepts that continue to influence stem cell research were being established (Reynolds 2018).

[9] Consider Rudolf Virchow's dictum: "All cells arise only from pre-existing cells." Of course, this does not answer the question of how the first cell came to be – a question deeply entwined with the origins of life itself. This view of cell division, although highly confirmed and deeply entrenched in current biological thought, emerged out of extensive efforts of discovery and debate (see the essays in Matlin et al. 2018). Schwann's own view, quickly disconfirmed, was that cells coalesce out of structureless cytoblastema in a manner analogous to crystallization or chemical precipitation out of solution.

[10] See, for example, Coleman (1977), Harris (2000), and Matlin et al. (2018).

[11] In sexually reproducing organisms, gamete cells can also fuse to form a new cell: a zygote. However, for reasons that become clear in the next section, stem cell phenomena are limited to divisions that occur within one organismal generation (one organism's lifetime). Thus the complexities of meiosis can be put aside here.

[12] "[M]ore than any other part, the membrane defines the cell, sets its outer boundary, and determines how the cell as an individualized unit interacts with its environment.... The membrane binds the cell into a single entity" (Liu 2018, 209). Of course, signals from the environment pass through the cell membrane all the time (and vice versa); the membrane is not a barrier.

a membrane and that which it bounds. This criterion determines, for any case of interest, how many cells are present. When cells reproduce by division, the process is finished when the membranes pinch off to separate. With such clear conditions for existence and distinctness, cells are perhaps the most straightforwardly countable living things – paradigmatic biological individuals. There is a quite clear and unambiguous way of determining whether an entity belongs to the kind "cell," what sets a cell apart from its environment, and how many cells there are in a given context. So cells are arguably our most straightforward example of a biological individual: a living entity that can be counted, picked out from its environment, and distinguished from other individuals of its kind.[13] Due to these well-characterized general features and clear boundaries, "cell" is a (relatively) well-behaved scientific category (or natural kind, to use the philosopher's term).

Most readers will be familiar with the textbook image of the cell – an idealized model indicating a set of distinguishable parts (organelles) within an enclosing membrane. Such a model depicts shared anatomic features that function in the physiology and metabolism of a (eukaryotic) cell: mitochondria, ribosomes, Golgi apparatus, nucleoli, endoplasmic reticulum, and more.[14] Textbook images of "the plant cell" and "the animal cell" – and perhaps "the prokaryotic cell" – are also fairly well-known. But these familiar images are idealized, generic *models*. Cells "in the wild" are very diverse. As noted previously, the natural kind "cell" subdivides into many more finely individuated kinds: *cell types*. Cells comprising multicellular organisms (the "building blocks" of organisms) are classified into a number of basic cell types, such as neurons, muscle, and blood cells. These, in turn, further subdivide into more finely grained types of neurons, blood cells, and so on.

This background suggests that "the stem cell" is also a cell type – a kind into which some cells of multicellular organisms are classified. Moreover, there are many varieties of stem cells – embryonic, hematopoietic, neural, induced pluripotent, and so on. So we seem to have a generic cell type, "stem cell," subdivided into more specific sub-types, as is the case for neurons and other familiar kinds of cells in multicellular organisms. If so, then we would have

---

[13] There is a lively debate in philosophy of biology and beyond about what it takes to be a biological individual (e.g., Guay and Pradeu 2016).

[14] "[C]onsider a typical biology textbook drawing of a cell. In most texts a schematized cell is presented that contains a nucleus, a cell membrane, mitochondria, a Golgi body, endoplasmic reticulum and so on. In a botany text the schematized cell will contain chloroplasts and an outer cell wall, whilst in a zoology text it will not include these items. The cell is a model in a large group of inter-related models that enable us to understand the operations of all cells. The model is not a nerve cell, nor is it a muscle cell, nor a pancreatic cell, it stands for all of these" (Downes 1992, 145).

already a fairly clear picture of the stem cell concept: a well-behaved natural kind, like the more inclusive kind "cell," distinguished from other cell types by a defining set of traits. As noted, cell types are defined in terms of traits held by their individual members: cell morphology, biochemistry, metabolism, molecular surface markers, gene expression profile, and so on. However, stem cells are *not* conceptualized and individuated in this way.[15] Cells that are identified as stem cells, of course, have size and shape, gene expression patterns, and other cell traits. But those traits are not what make them stem cells. Stem cells are defined not (only) in terms of the traits they have but by what they can transform into. The concept *stem* does not merely qualify the category "cell" – it introduces another idea altogether.

Stems and Lineages

The term "stem", in ordinary life, is both noun and verb. A *stem* (noun) is part of a plant, the stalk that supports a flower. To *stem* from something (verb) is to originate from that thing. Both meanings are relevant to the concept of stem (or stemness) applied to cells. A stem is a site of growth – an active source that supports or gives rise to something else. That other thing is, in a sense, primary: a *stem* (noun) plays a supporting role; the object of *stemming* (verb) is that which comes from some origin or source. A stem, conceptualized as a noun or verb, entity or process, is so because of the thing it supports or gives rise to. A stem (or stemming) is powerful, important, and potent – but only relative to something else, not in itself. It is temporally prior but conceptually secondary, to that which it gives rise to. The term *stem cell* (Stammzelle) was coined with exactly these connotations in mind. Ernst Haeckel introduced it in 1868, one of a spate of neologisms associated with his ideas about the Tree of Life (see the next section). The Stammzelle, for Haeckel, referred to the hypothetical origin of all the diverse kinds of cells and organisms – the origin of growth for the phylogenetic tree uniting all living things. Although the meaning of "stem cell" today is not the same as Haeckel's, his original connotations remain. So the concept of a stem cell is somewhat peculiar, combining two very different ideas. A *cell*, as we have seen, is a well-characterized biological entity, observable via relatively simple technology and clearly distinguished from its environment and other cells by a bounding membrane. A *stem*, in contrast, is the beginning of a process, the point of origin for something that is to be.

This internal tension has important scientific consequences, explored in later sections. One way to manage the tension, reconciling stem cells' dual aspect, involves another piece of conceptual background: the idea of a *lineage*. In

---

[15] The argument for this unfolds across the next two sections.

biology textbooks, a "lineage" is defined as a sequence of biological entities (e.g., organisms, species, cells) linked by ancestor-descendant relations. Only entities that reproduce can form lineages, each an unbroken chain of descent. So the concept of a lineage presupposes some process of reproduction. Boundaries of a lineage are fixed by the chain of inheritance across generations, beginning with an originating ancestor and ending with the last descendants. The association with the notion of a stem, an origin of growth, is obvious. Fittingly, lineages are often depicted as "tree diagrams" that represent generational ties between reproducing entities. Humans are often interested in genealogical relationships, so tree diagrams (in the form of pedigrees) are a familiar representational form. These everyday lineages are of individual persons, each occupying a distinct position in the family tree. In philosophy of science, the most familiar lineages are not of individual persons or organisms but of species: Darwin's famous sketches of phylogenetic relationships that connect all the diverse forms of living things into a Tree of Life. The nature and definition of biological species is perennially unsettled. The lineage form presumes only that they can stand in ancestor-descendant relations to one another (i.e., that some process of reproduction is involved). In many cases, the process of interest is the reproduction of organisms in populations, which in turn involves the transmission of genes from parent to offspring. Philosophers of science have primarily discussed lineages of this sort, seeking to articulate further constraints that define "units of evolution." Neto (2019) takes a broader view, arguing that there are multiple kinds of lineages, each corresponding to concepts and practices of different biological research programs (evolutionary, developmental, phylogenetic, etc.). Lineages in stem cell biology tend to support such pluralism.

Any lineage whatsoever can be characterized in terms of its "tree structure" – that is, by features of the graph connecting the point(s) of origin to the most recent descendants (Figure 1). A tree structure consists of

- origin points (no incoming edges);
- end points (no outgoing edges); and
- branch points (one incoming edge, two or more outgoing edges).

A lineage extends from one or more ancestors, through lines of reproductive descent, to one or more end points. The latter may be temporary (the newest generation) or permanent (no more reproduction; the line has ended). The number of origin and end points establishes the boundaries of a lineage. The structure between them consists of some number of generations (e.g., the "depth" of the tree). The simplest such structure is just a linear sequence of entities with no branching. Depending on the manner of reproduction and the entity undergoing it, branch points may be bifurcations or trifurcations; thicker

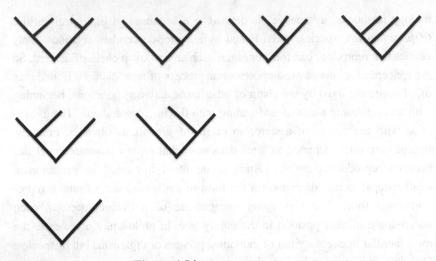

Figure 1 Lineage tree structures

"bursts" are rare. A *cell* lineage consists of successive cell generations, organized by reproductive relations. Because cells reproduce by binary division, cell lineages have a bifurcating tree structure. So the idea of a cell lineage dovetails neatly with tenets of cell theory:

- Cells reproduce by binary division; a parent cell divides to produce two offspring cells.
- Generations of cells linked by reproductive division form a lineage.

Fields that study lineages are interested in not only the pattern of ancestor-descendant chains but also the pattern of variation among the entities so related. For evolutionary biology and phylogenetics, genetic or genomic variation is often of primary importance. Not so for studies of organismal development, however. The lineages of most relevance to organismal development are cell lineages, which exhibit very little genetic/genomic variation. That's because all the cells comprising a developmental lineage are descended from a single ancestral cell: the zygote.[16] Although some genetic/genomic variability arises over the course of the many cell divisions involved in organismal development, for the most part descendant cells are genetic copies of the original. Of more interest is *epigenetic* variation among cells: changes in gene expression that produce changes in cell structure and function. A key goal of developmental

---

[16] Unicellular organisms form lineages through this same reproductive process (cell division). These lineages are at once cell and organism lineages. In multicellular organisms, cell and organism lineages do not coincide in this way. Every living thing originates with a single cell (this includes viruses, albeit in a different way).

biology is to describe the sequence of changes leading to specialized cell types that make up an organism's body.

A Note on Definitions

This section has introduced the conceptual background of stem cell research: interrelated notions of cell, differentiation, cell type, and lineage. The dual nature of stem cells, as clearly bounded biological individuals and protean origins of growth, is reconciled in the figure of a lineage. This gives us a first pass at defining "the stem cell" – as the origin point of a cell lineage. Although on the right track (as the next section argues), this epigrammatic answer is too abstract to fully characterize the stem cell concept as it figures in scientific practice today. This brings us back to the opening question: What is a stem cell? The next section proposes an answer.

As a preliminary to that discussion, a brief note on definitions is needed. Many philosophers are accustomed to definitions as the cornerstone of conceptual analysis – necessary and sufficient conditions for a term or concept to apply. Philosophers tend to prize sparse analytic elegance, a kind of pristine rigor, as the mark of excellence. Many areas of biology do not share this epistemic/aesthetic value, and their products do not exhibit it. Stem cell researchers, like most scientists, do not offer definitions in the way of analytic philosophy. They do introduce new terms, communicate with the general public, and teach novice scientists the meaning of the term "stem cell". All these activities involve a definition, in the working scientist's sense. And these working scientific definitions go some way toward clarifying the concept of *stem cell* – but not far enough to be philosophically satisfactory. The eponymous focal concept of stem cell biology is never explicitly defined in a way that directly reflects that concept's role in scientific practice.[17] Instead, scientists' explicit definitions of "stem cell" are for outsiders and novices. They are therefore simple and abstract, like the textbook model of a generic cell. Hereafter, when I refer to "definitions" of the term "stem cell," I mean, roughly, a conceptual model of that idea. Clarifying the stem cell concept amounts to offering a more nuanced and detailed model of the stem cell concept, and thus offers a clearer window on the practices of stem cell research than the thinner definitions offered by scientists.[18] So the goal of the next section is to bridge the gap between the thin definitions offered by scientists to nonexperts, and the stem cell concept as it figures in scientific practice.

---

[17] This may be typical of experimental sciences, though that issue is beyond the scope of this short Element.

[18] Thanks to an anonymous reviewer for pressing me to clarify this point.

2 Stem Cells: The Very Idea

The previous section sketched the dual nature of stem cells, as reflected in their name: clearly individuated biological entities (cells) and generative "springs of life" that give rise to other things (stems). The concept of a cell lineage, with the stem cell as its origin point, encompasses this dual character. But that epigrammatic characterization does not capture the stem cell concept in scientific practice. The next logical step is to look at how stem cells are defined by the researchers who study them. And here we hit a snag. The question "What is a stem cell?" does not preoccupy scientists.[19] They are more concerned with the topics of the next two sections: finding and using stem cells. But the ways they pursue these activities are largely impenetrable to outsiders.[20] Stem cell biology is a complex, untidy, technically challenging, unfinished, and enormously innovative area of science. To engage with it conceptually, without investing in years of training, requires some kind of orienting framework. Lacking such, one is swept away by a sea of technical details, rapidly changing jargon, and unstated assumptions that invite misunderstanding. The idea of a stem cell, being conceptually central for stem cell research, is the natural framework to use. But it will take some philosophical work to articulate it. That is the task of this section.

Of course, this philosophical clarification should be grounded on stem cell science. I begin by dispelling some associations that have accreted into a popular notion of stem cells. After this ground-clearing, I look to the term's historical origin: the German "Stammzelle." The current concept is not the same as the original, yet some key ideas remain. I then turn to present-day definitions of "stem cell" offered by scientists to nonexperts. From this grounding in past and present scientific practice, I articulate a philosopher's definition of "stem cell" – a conceptual anatomy of the concept, which offers a window into stem cell research.

Beyond Embryos

Most readers are probably familiar with the idea of a stem cell, at least in broad outline. In everyday use, the term is closely associated with two other ideas: medical promise and embryo destruction. Both are intensely value-laden, in somewhat opposed ways. On the one hand, stem cells are tokens of medical promise and hope; potential means for curing injuries and diseases ranging from

---

[19] Dröscher (2014) argues that, historically, the meaning of "stem cell" is largely tacit, transmitted via metaphor and diagrams (primarily cell lineage diagrams, discussed in this section). This goes for present-day stem cell research too.

[20] Stem cell research is not unique in this respect. Many fields that are experiment-driven have this character, which contributes to the neglect of experiment by philosophers.

Parkinson's to macular degeneration to male pattern baldness. Their apparently limitless regenerative powers make for an ever-renewable source of healthy cells and tissues.[21] Destruction of human embryos is the negative shadow of these bright possibilities. Although the science goes back much further, the idea of a stem cell entered public awareness with the announcement of cultured human embryonic stem cells (Thomson et al 1998). These "immortal" cells, continuously propagating in artificial culture from that day to this, were derived from early-stage human embryos donated from IVF clinics.[22] The idea that stem cells' restorative powers are bought at the cost of a new human life is evocative, posing a tradeoff problem that captivates some bioethicists. Public and philosophical discussion of stem cells understandably focuses on ethical and policy issues surrounding the use of human embryos in research.[23]

However, the ethics and policy approach does not fully capture the scientific idea of a stem cell. For one thing, most stem cell research does not involve the use of embryos, human or otherwise. Indeed, before 1998, the term "stem cell" usually referred to a mammalian bone marrow cell capable of producing the full range of blood and immune cells, the ancestral hematopoietic ("blood-making") cell. Yet the association between early embryos and stem cells goes back to the latter term's origin. Overall, "stem cell" has a complex, historically shifting meaning, of which embryos are only a part. Second, although medical applications are important in motivating and guiding stem cell research, the field today is dominated by basic science and preclinical research. With the exception of bone marrow transplantation therapy, used for decades to treat leukemia and other blood disorders, medical applications of stem cells remain experimental or unconfirmed.[24] Both medical applications and embryos are important parts of stem cell research, but neither gets to the core of the stem cell concept. To reach that core, we need to look closely at how the term "stem cell" is defined in scientific practice.

Historical Origins

The term *stem cell* (Stammzelle) was introduced in 1868 by Ernst Haeckel.[25] He did not, however, define it explicitly, but used the term "stem" ("Stamm") in

---

[21] It is this promise that sustains stem cell clinics, peddling cures for nearly any ailment one can think of, unencumbered by scientific evidence or regulatory approval. (See Section 4.)

[22] These were "extra" embryos from patients' in vitro fertilization efforts. See Franklin (2013) for an erudite discussion of IVF as a source of stem cells.

[23] Literature on this topic is voluminous. See, for example, Blasimme et al. (2013), Franklin (2013), Hauskeller et al. (2019), Maienschein (2003), Waldby and Cooper (2010).

[24] See Section 4.

[25] Haeckel was a prolific wordsmith for biology, coining the terms "ecology" and "phylogeny" – and, not incidentally, the phrase "ontogeny recapitulates phylogeny."

various ways associated with his famous Tree of Life image. The Tree of Life depicts evolutionary history and relations among taxonomic groups. Haeckel's vision of this was highly speculative, and his usage of "Stammzelle" no exception: "the different primeval 'original cells' [Stammzellen] out of which the few different main groups or tribes [Stämme] have developed, only acquired their differences after a time, and were descended from a common 'primeval cell' [Urstammzelle]" (Haeckel 1876, II, 41).

The original meaning of "stem cell", then, referred to the single-celled base of the genealogical tree of all living things – the primordial stem from which grew all the diverse forms of life we see today. Haeckel further theorized that organisms' development and species' evolution are intertwined. The phrase "ontogeny recapitulates phylogeny" – which he also coined – neatly epitomizes this theory. Stated more plainly, the development of an individual organism reflects the phylogenetic history of species relationships. In ontogeny (that is, an organism's development), the stem cell is the starting point of the entire process:

> I have given a special name to the new cell from which the child develops, and which is generally loosely called "the fertilized ovum" or "the first segmentation sphere". I call it "the stem-cell" ... because all the other cells of the body are derived from it, and because it is, in the strictest sense, the stem-father and stem-mother of all the countless generations of cells of which the multicellular organism is to be composed. (Haeckel 1905, I, 130–131)

Haeckel's idea of a stem cell included the concepts of cell and cell lineage, as well as ideas about cell reproduction and the organism. The stem cell is the ancestor of all cells of an organism's body, the latter formed by many generations of cell reproduction. The stem cell, or zygote, is the repository of *developmental potential* for the whole organism.

The meaning of "stem cell" today is rather different, for at least three reasons. First, the processes of development and of evolution are conceptualized separately – to such an extent that there is an entire sub-field dedicated to re-integrating them (evolutionary developmental biology). Stem cells are firmly on the development side of that divide. Evolutionary biology today does not include the idea of a stem cell. (Lineages, however ...) Second, the fertilized egg, or zygote, is not usually considered a stem cell today, while many other cells in a developing organism are (along with others outside organisms; ex vivo stem cells). Third, theories of organismal development today are quite different from Haeckel's day. Yet some continuity with Haeckel's original idea remains. Tracing these links, Ariane Dröscher (2012, 2014) argues that our current stem cell concept descends from a combination of Haeckel's term and August

Weismann's theory of inheritance. Although long since disconfirmed in its particulars, Weismann's theory remains in the background of much biological thought. Weismann theorized that organismal development begins with a single cell (the "Urzelle") "containing the generative power of the entire future organism" (Dröscher 2014, 164). He was interested not only in one organism's development but how the developmental process recurs across generations of organisms. How is developmental potential transmitted? Today, we have detailed and precise (though still incomplete) molecular genetic explanations of this process. Weismann's was a pioneering theory, far ahead of any molecular details or what we consider genetics today. One of his key ideas was a fundamental distinction between two kinds of cell lineages within an organism: "the germ line" and "the soma" (Figure 2). The former crosses generations, while the latter is an organism's body, lasting its lifetime but no more. Another key idea was that cell division works differently in germ-line and somatic lineages; they are distinguished by different mechanisms of cell reproduction. The Urzelle is the origin of both, but soon after, an embryo's cells are irrevocably split into germ line and soma. (The idea that cell development within an organism proceeds by way of "forking paths" or bifurcating lineage pathways, remains influential in stem cell biology today – though the two ideas noted here are consigned to the history of science.)

Weismann's germ-line/soma distinction, coupled with his ideas on cell division, outlines a cell-based theory of organismal development. Briefly, the Urzelle's generative potential is fragmented through cell division – but only in the somatic lineage. Somatic cell divisions give rise to (nearly) all an organism's cells, splitting up the Urzelle's potential among diverse cell lineages. This "unequal segregation of the germ-plasm" produces all the different cell types that make up an organism's body, which in turn compose the tissues and organs that sustain its life. That is, in Weismann's theory, cell division is the underlying cause of an organism's development, from a single starting cell to

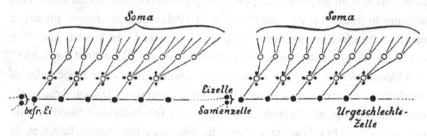

Figure 2 Diagram of Weismann's germ/soma distinction. From Boveri, *Über die Entstehung* ... (1892), detail of Figure 1, p.118.

a complex body made up of many cell types, tissues and organs. The germ line is held back from all this, however, sequestered from the rest of an organism's development. This is also because of cell division: in the germ line, cell division preserves the Urzelle's full developmental potential. This lineage produces only the reproductive cells that contribute to the next generation of organisms ("continuity of the germ line"). Unlike cells of the somatic lineage, germ-line cells do not perform functions in the organism's body; their role is just to form the next organismal generation. This requires that they retain full developmental potential – and this in turn requires that they not participate in the process of somatic development. This idea – that developmental potential is diminished through the actual process of development – remains influential in stem cell biology today; part of biology's conceptual inheritance from Weismann.

Because Weismann theorized cell division as the primary mechanism of organismal development, his cell lineage tree diagrams amount to a causal explanation of development. Though Weismann did not adopt Haeckel's term, his Urzelle is very like a Stammzelle: "a single cell representing the starting point of ontogenesis . . . a sort of ancestral remainder possessing maximum prospective potency" (Dröscher 2014, 165). Weismann's cell lineage diagrams posit an Urzelle as the unique start of an organism's development. Branch-points from this "primordial cell" represent cell division events: the primary mechanism of cell and organismal development. Other embryologists, including Theodor Boveri, Oskar Hertwig, and E. B. Wilson, soon took the natural step of combining Haeckel's term with Weismann's cell tree diagrams. In late-nineteenth- and early-twentieth-century embryology, the term "stem cell", though rarely explicitly defined, referred to cells possessing the full complement of inherited material, understood as containing the full developmental potential for the whole organism.

Over ensuing decades, this early stem cell concept faded from use. However, the image of a cell lineage tree emanating from a stem cell found an enduring home in hematology, the study of blood (Maehle 2011, Dröscher 2014). For most of the twentieth century, the stem cell concept was localized to pedigrees of blood cells. Blood is an atypical tissue, circulating throughout our bodies rather than working in place. Yet its associations with vitality and inheritance preserved many of the older connotations of the term "stem cell." In humans and other mammals, blood is composed of fluid serum and cells, the latter "red" or "white." Red blood cells do the work of oxygen transport, while white blood cells, of which there are many types, mediate the immune response. Though these circulatory functions persist throughout an organism's life, individual blood cells do not. A single red blood cell, in humans, circulates for a few months, then dies, its parts fragmenting, to be reused or excreted. The life-span of white blood cells is more variable, ranging from weeks to years, but the end is the same. The life cycles of organisms and blood cells are not synchronized. For an organism to have a long and healthy life, new blood cells must continually develop to replace those that die off. The vital functions of blood require regeneration on a massive scale; about one trillion (10^{12}) new blood cells per day in adult humans. Researchers began attempting to trace the pathways by which blood cells develop in the 1870s. By that time, microscopic study had revealed different types of blood cells, based on morphology. The anatomical location of blood stem cells was also known: bone marrow (in adult humans). So a general lineage picture of blood cell development was in place more than a century ago.[26] The hypothesized origin point is a stem cell localized to bone marrow, the end points the known blood cell types. But stages linking the two – branch-points – are unclear. Researchers offered hypotheses about those relations (i.e., the pathways of blood cell development, represented as cell lineage tree diagrams).[27]

In blood cell lineage diagrams, unlike their Weismannian antecedents, cell divisions are not depicted. Instead of generations produced by cell division, the lineage tree is organized by morphologically distinct kinds of cell: cell types, arranged in a "differentiation hierarchy." For Weismann, of course, cell division was the mechanism of cell and organismal development – but with the demise of that theory, cell reproduction and cell development are uncoupled. The uncoupling is not total; there must be cell division to generate enough cells to build the various parts of an organism, from a single-cell starting point. But the changes in cells (epigenetic changes; see section 1) that lead to all the specialized cell types comprising an organism's various tissues and organs do not proceed in lockstep with cell division. So the bifurcating structure of cell reproduction does not dictate the structure of cell lineage trees representing development. For blood, the structure is delimited by the originating stem cell and diverse branches corresponding to distinct lineages of blood and immune cells: red and white, B and T, helper and killer T cells, and so on. But there are many ways to fill in the intermediate stages; many different arrangements of branch points are possible. These branch points represent intermediate cell types, neither stem cells nor fully specialized (a.k.a. differentiated). To locate these stages in the blood cell hierarchy, intermediates needed to be distinguished from other cell types in the blood cell lineage, by some set of characteristic cell traits (size, shape, biochemical profile, surface receptor profile, gene expression pattern, etc.). In the early twentieth century, the distinguishing traits that organized a blood cell differentiation hierarchy were primarily morphological, easily detected under a simple light microscope. As more methods for characterizing cells were innovated and

---

[26] Maehle (2011) gives an informative history of this strand of research.

[27] Another example is Maximow's hypothesis of blood cell hierarchy based on morphology of cells found in bone marrow (1909).

refined, hematologists were able to distinguish among more and more different types of blood cells, and thus to construct increasingly fine-grained hierarchies. Because white blood cells mediate the immune system, hematological studies of blood cell development merged with studies of cellular immunology. Representing cell development as a lineage tree ("differentiation hierarchy") remains an important theme in stem cell research today.

Detailed examination of the history of stem cell research is beyond the scope of this essay.[28] Instead, I will summarize and skip ahead. Until the late twentieth century, the idea of a stem cell figured mainly in the study of blood (hematology). This changed in 1998, when the innovation of cultured human embryonic stem cells brought the idea into the public eye, with the attendant hopes, fears, and ethical tradeoffs discussed previously. So, what is the scientific meaning of "stem cell" today?

Scientific Definitions

As noted earlier, working scientists rarely define "stem cell" explicitly. Instead, they report a great many detailed cellular, molecular, and genetic characterizations of *specific varieties* of stem cells. We have already encountered two important ones – embryonic and hematopoietic. But there are dozens if not hundreds of others (depending on how one distinguishes varieties). There is no single classification system for stem cells, nor any proposed system that covers all stem cells. Some varieties serve as important research foci: induced pluripotent, muscle, neural, and trophoblast stem cells, and many others. An influential textbook, *Essentials of Stem Cell Biology* (Lanza et al 2009), indexes thirty-eight different varieties of stem cell: adult, amniotic/amniotic fluid, bone marrow, cancer, cardiac, cord blood hemato-poietic, dental pulp, embryonic, embryonic germ, embryonic kidney, embryonal carcinoma, epiblast, epidermal, (hair) follicle, germline, hematopoietic, induced pluripotent, intestinal, keratinocyte, leukemic, liver, mesenchymal, multipotent, multipotent adult progenitor, myogenic, neural, pancreatic, pancreatic liver, pluri-potent, post-natal, renal, skeletal, skeletal muscle, solid tumor cancer, somatic, tongue, trophoblast, and very small embryonic-like. (This is an underestimate of the number of distinct stem cell varieties circa 2009, as it leaves out different species designations and "progenitor" cell types that are not explicitly labeled as stem cells.[29]) Nor does this textbook list suggest any clear taxonomy of stem cell varieties; different designations sometimes overlap or use cross-cutting criteria. In this, the textbook reflects the field as a whole; there is no single principle for

---

[28] See, for example, Fagan (2007), Kraft (2009), Maehle (2011), and Maienschein (2003).

[29] Sometimes progenitors are classified as distinct from stem cells; sometimes the two are lumped together. This adds another layer of variability to characterizations of "stem cell."

distinguishing stem cell varieties. Their diverse monikers reflect this. Many stem cell varieties are named for the part of an organismal body they give rise to (e.g., myogenic, neural). Others refer to the location they are found within an organism's body (e.g., amniotic), the stage of organismal development at which they can be found (e.g., embryonic), how they are made (e.g., induced), associated pathologies (e.g., leukemic), or generic differentiation potential (e.g., multipotent). Overall, a look at the science reveals many varieties of stem cell, each with its more- or less-detailed scientific characterization, but no clear classification criteria.

A few commentators have noted that the terminology of stem cell research is strikingly confusing and disordered (e.g., Rao, 2004). But this does not seem to trouble most scientists working on stem cells. To my knowledge, there has been no attempt by stem cell biologists to systematically classify all the various types of stem cells.[30] The issue is not raised as a challenge at major meetings, in calls for funding, special journal issues, or the leadership of the International Society for Stem Cell Research (ISSCR), the field's premier professional society. In practice, stem cell researchers focus on whatever specific varieties of stem cell their laboratories are equipped to handle and do not concern themselves with the big picture of stem cell classification. This makes it challenging to sketch what the big picture looks like for stem cell research – if there is a single such picture at all.

But scientists have to learn about stem cells somehow. No stem cell researcher begins their career steeped in the experimental lore and intricate characterization of any stem cell variety. One way scientists communicate the basic idea of a stem cell is epitomized in the phrase: "asymmetric cell division in a niche." Stated more fully, this working scientific characterization defines a stem cell by the ability to divide asymmetrically, that is, to produce one offspring that is a stem cell and the other a more specialized cell, in some extracellular environment (Figure 3). This definition often appears in introductory or educational materials for scientists-in-training, and in presentations to scientists outside stem cell research. For a broader audience (including most philosophers), more explanation is needed. Recall that cell division is the mechanism of cell reproduction: a parent cell divides to produce two offspring cells.[31] By convention, if the two offspring

---

[30] There are proposals for methods of unambiguous nomenclature for human pluripotent stem cell lines (e.g., Kurtz et al., 2018). This would be a finer grain of classification than the varieties noted here, limited to stem cells in one species with early embryonic developmental potential – not the entire range of stem cells.

[31] The above is mitosis. There is also meiosis, a special case of cell division coupled to sexual reproduction: generation of a new organism from the union of two germ cells. Stem cell phenomena do not include meiosis, so I set it aside here. One further note: though biologists often refer to "mother" and "daughter" cells, I avoid these gendered terms. Cell biology is complex enough without piling on the additional baggage of myriad assumptions and biases associated with gender divisions in society. Women in science (and philosophy) who struggle

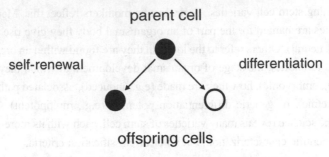

Figure 3 Diagram of asymmetric cell division

cells resemble one another, then the division event that produced them is termed "symmetric"; otherwise, it is "asymmetric."[32] A cell division event that produces a stem cell and a more differentiated cell is obviously asymmetric. Contra Weismann's theory, the *cell niche* drives cell development, not the internal mechanism of cell division. The cell niche consists of features of a cell's immediate environment that influence the developmental trajectory of that cell and its lineage descendants. It includes physical and chemical factors, other cells and their surface and secreted products, and noncellular structures such as the extracellular matrix. Offspring produced by a cell division event occupy somewhat different niches, simply in virtue of the fact that there are two cells where before there was one. One offspring cell occupies the same niche as the parent – and so it is a stem cell. The other occupies a niche that induces it to differentiate along some pathway or other. This offspring cell is not a stem cell but more specialized, displaced from the origin point of a cell lineage and moved to a position further along some developmental pathway. A cell division event satisfying the working scientific definition neatly encapsulates the reproductive and developmental processes that distinguish stem cells from other biological entities. Self-renewal and differentiation are illustrated in a single cell division event. Such simplicity is rare in stem cell research. So it makes sense that the concept of "asymmetric cell division in a niche" is often used to introduce scientists outside stem cell research to its basic ideas. However, that simple introductory definition does not map onto the idea of stem cells in research practice.[33] To get at the latter, we need first to

---

with obstacles created by entrenched assumptions and biases have good reason to be suspicious of gendered language, especially where it is unnecessary.

[32] The onset of cell development in early embryos or models thereof is sometimes referred to as "symmetry breaking."

[33] So this definition is a simple model, intended for use by other scientists – it captures the basic idea of a stem cell in terms of processes familiar to most biologists (cell division, extracellular location, the trajectory of development).

pinpoint the most basic, minimal idea of a stem cell in scientific practice today – the bare essentials, so to speak.

Public-Facing Definition

The simplest and most general characterization of stem cells currently on offer is the public-facing definition, used by stem cell scientists communicating with the wider public. (It also appears in scientific contexts, usually in introductory sections of textbooks, articles, and presentations.) For example, the International Society for Stem Cell Research and the European Stem Cell Network, respectively, offer the following in their outreach materials:[34]

> Stem cells: Cells that have both the capacity to self-renew (make more stem cells by cell division) as well as to differentiate into mature, specialized cells.

> Stem cell: A cell that can continuously produce unaltered daughters and also has the ability to produce daughter cells that have different, more restricted properties.

Similarly, the first issue of *Cell Stem Cell* (2007) defined the eponymous entity as a cell with "the capacity to both self-renew and give rise to differentiated cells" (Ramelho-Santos and Willenbring, 2007, 35). The most recent edition of *Essentials of Stem Cell Biology* (2013) states that "stem cells are functionally defined as having the capacity to self-renew and the ability to generate differentiated cells" (Melton, 2013, 7).[35]

These prominent examples (and there are many others) all converge on two abilities as essential to being a stem cell.[36] Stem cells can (1) divide to produce offspring that are also stem cells (self-renewal), and (2) generate descendant cells specialized for particular functions within a developed organism (differentiation). Stem cells are thus defined purely in terms of their reproductive and developmental abilities – that is, in terms of what they *do*, not traits they have (such as structure, morphology, physiology, gene expression, etc.). A clearer picture of the conceptual core of stem cell research emerges from a closer examination of these two defining abilities. Self-renewal is a reproductive process; differentiation a developmental one. So stem cells are functionally

---

[34] Both circa 2016. See also the Cell Therapy and Regenerative Medicine Glossary (2012), National Institutes of Health Stem Cell Glossary (2015), and further references in Fagan (2013a).

[35] Melton goes on to expand on the general definition: "a more complete functional definition of a stem cell includes a description of its replication capacity and potency ... a working definition of a stem cell line is a clonal, self-renewing cell population that is multipotent and thus can generate several differentiated cell types" (2013, 7–9). This more elaborate definition, intended for a scientific audience, retains self-renewal and differentiation as the defining capacities of stem cells.

[36] The same abilities are part of the working scientific definition, just not as explicit (as discussed later).

defined in terms of two central biological processes – reproduction and development – conceptualized at the cellular level. Stem cell researchers do not explicitly define these processes, even in introductory treatments.[37] Thus what follows is my own philosophical clarification.

"Self-renewal" refers to stem cells' reproductive ability; stem cells can beget more stem cells. The ability to self-renew is realized (made actual) by cell division. For individual cells, the term is somewhat misleading, because stem cells do not literally generate themselves through cell division. Instead, a parent produces one or two offspring stem cells of the same variety; parent and offspring are different cell-individuals but the same kind of stem cell.[38] So self-renewal implies a cell lineage: individual cells related by reproduction. A self-renewing stem cell gives rise to a lineage of stem cells, extending two or more generations. (Asymmetric cell division, as discussed previously, is a minimum case.) In practice, this means that some of a stem cell's lineal descendants (actual or potential) exhibit the features that make it a stem cell of whatever variety it is. Self-renewal is a peculiarly recursive concept, presupposing a set of character traits associated with each stem cell variety. It is also a temporal notion, involving some number of cell division events or a time interval over which these occur. These considerations suggest the following analysis:

> A cell is capable of self-renewal just in case it initiates a cell lineage produced
> by (at least) n cell divisions such that parents and offspring are the same with
> respect to some set of characters C.

The characters in C are just those taken to define a stem cell of some particular variety.[39] Whatever is included in C, self-renewal requires that these characters remain constant across some number of cell generations. The latter represents the extent of a stem cell's self-renewal ability (sometimes unlimited, others with an upper limit).

The ability to differentiate is, roughly, the ability to undergo a cell-level process of development. The simplest way to conceive this process is as a linear sequence of cell types (stages) ordered in time, analogous to whole-organism stages of embryonic development. (The latter is an enormously influential mode of representation in embryology and its successor discipline, developmental biology;

---

[37] This is one reason that learning about stem cells is notoriously difficult; there is not much in-between the details of cutting-edge research practice and the short glossary definitions offered to the public. To address this gap, the ISSCR is currently developing a standard syllabus for teaching students the basics of stem cells (ISSCR 2020).

[38] Stem cells are self-renewing at the population level. The term's scientific usage dates back to when this was the finest grain possible for their experimental detection (see Section 3).

[39] These vary widely, depending on the variety of stem cell, and additional characters are added over time, as researchers learn more about stem cells of a given variety (see Section 3).

Hopwood 2005). However, there is one very important disanalogy between organism-level development and cell-level development within a multicellular organism. While the organism remains the same biological individual throughout the developmental process, this is not the case for cell-level development. Like self-renewal, differentiation involves a cell lineage. Also like self-renewal, the lineage requires cell division and takes the form of branching pathways (Figure 4). However, branch-points in a differentiation hierarchy do not represent cell division events. The one-way developmental process provides a basis for ordering cell types as more or less specialized. But this developmental ordering is uncoupled from the process of cell division.[40] Instead, branch-points represent alternative developmental pathways that cells may follow. So the cell-level developmental process is not linear, but branching, representing the various pathways leading to specialized cell types that make up the body of a fully developed organism. Each pathway consists of a sequence of stages of cell types, leading from the original undifferentiated stage to the fully specialized, terminally differentiated stage. Cells are more or less differentiated according to their position in this branching lineage tree structure. Each pathway ends with

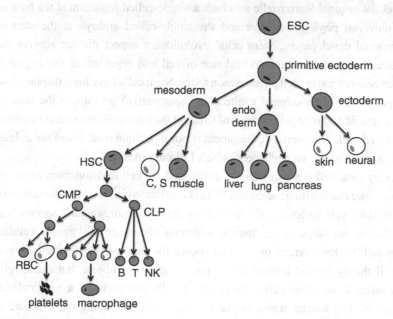

Figure 4 Schematic diagram of mammalian cell developmental pathways

---

[40] An individual cell can transform from one cell type to another, while cells of one type can undergo cell division.

a "mature" (i.e., fully developed) cell type that contributes to the body of an adult organism. In abstract terms:

> A cell is capable of differentiation just in case that cell initiates a cell lineage *L*, end points of which correspond to more specialized cell types.

Again, asymmetric cell division is a minimal case: a developmental cell lineage tree consisting of stem and endpoint stages only. So this conceptual analysis encompasses both the working scientific and public-facing definitions of stem cells. More is needed, however, to connect this clarification of self-renewal and differentiation with stem cell research practice.

Stem Cells and Organisms

So far, I have shown that explicit scientific and public-facing definitions of "stem cell" converge on two abilities: self-renewal and differentiation (i.e., cell-level reproduction and development, respectively). I have clarified the standard scientific meaning of both processes. However, although self-renewal and differentiation are cell-level processes, the idea of a stem cell is not a purely cell-level idea. Multicellular organisms are implicated in at least four ways. The first is historical: Haeckel's original Stammzelle was both a single-celled organism at the base of the universal phylogenetic tree and the single-celled embryo at the start of organismal development. Stem cells' evolutionary aspect did not survive the nineteenth century, and the coincidence of cell and organism in the zygote is not considered a stem cell phenomenon today. Stem cells have been displaced, so to speak, from their original exalted (and speculative) position at the base of phylogenetic and ontogenetic trees of life. Yet they are still tree-making entities, entangled with organismal development if not its unique root. There are at least three aspects of this entanglement, which I discuss in turn.

Every stem cell is a part of, or is descended via cell division from a part of, exactly one multicellular organism.[41] This is reflected in the scientific names of many stem cell varieties, which refer to the source organism's species and developmental stage (e.g., "human embryonic stem cell," "mouse epiblast stem cell"). Moreover, an organismal source for stem cells fits with the tenets of cell theory in that a stem cell is not an *ab initio* object, but necessarily descended from other cells. Those other cells are parts of a multicellular organism. The source organism plays a large role in shaping the structure of lineages associated with a stem cell. Three features are particularly impactful in this regard:

---

[41] And every organism – every living thing, arguably – begins as a single cell; today's stem cell concept thus inverts and complements Haeckel's original notion.

- species (e.g., mouse, human);[42]
- developmental stage (e.g., five-day embryo, gastrula, adult organism); and
- location within the organism (e.g., inner cell mass, neural crest, bone marrow).

The first two are features of an organism, while the third specifies a relation between cell and organism (i.e., that the former occupies a particular position within the latter). An incisive way to express their influence is that stem cell capacities are *context-dependent*, being derived from a particular site within a multicellular organism of a particular species and developmental stage.

Relatedly, a stem cell's potential is roughly correlated with the source organism's developmental stage: the earlier the latter, the greater the former. As an organism's development proceeds, the developmental potential of stem cells derived from that organism diminishes (Figure 5). The maximum potential is *totipotency*: the capacity to produce an entire organism (and, in mammals, extra-embryonic tissues). In animals, totipotency is limited to the fertilized egg (the original Stammzelle) and products of early cell divisions. The maximum developmental potential for stem cells as the term is used today is *pluripotency*: the ability to produce all cell types of an adult organism. More restricted stem cells are *multipotent*: able to produce some, but not all, mature cell types. Stem cells that can give rise to only a few mature cell types are *oligopotent*. *Unipotency* is minimum potential: the ability to produce only a single cell type. This terminological ordering of potencies offers a generic, imprecise way of classifying stem cells. As a general rule, stem cells derived from an early embryo are pluripotent, while stem cells isolated from parts of a fully developed organism of the same species are (usually)

Organismal development

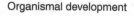
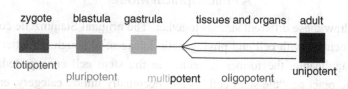

Figure 5 Correlation of cell and organismal development

---

[42] Most stem cell research focuses on animals rather than plants, though both undergo development (and stem cells in *Arabidopsis* are an important research area). This book accordingly deals for the most part with stem cells in animals. Any full explanation of development, however, must extend to plants as well.

multi-, oligo-, or unipotent. Whether this rule is strict and exact for all organs and tissues, or pluripotent stem cells can be reliably found in adult organisms, is a long-running controversy (see Fagan, 2013a, ch. 3). But the general pattern appears to hold, at least roughly: "[e]mbryonic development and the subsequent adult life are viewed as a continuum of decreasing potencies" (Can 2008, 59). In this idea, at least, Weismann's theory of development retains a hold in current biology.[43]

Finally, and most subtly, organisms are implicated in the concept of a stem cell's developmental potential. As noted, each variety of stem cell has the ability to produce some range of more differentiated cell types. These ranges are expressed as general categories of stem cell potential, or potency: pluripotent, multipotent, and so forth. If stem cell potency were characterized in cellular terms only, the "pluri-", "multi-", and "uni-" monikers would simply refer to raw numbers of cell types. But this is not the case – as the remarks provided reveal. The categories of potency do not refer to cell types considered in isolation, so to speak, but to their contributions to organismal development. Pluri-, multi-, and unipotent stem cells are capable of giving rise to all, some, or one of the cell types making up a multicellular organism (of the same species as their source). Naming conventions for stem cell varieties also reflect the import- ance of organisms as developmental products of stem cell capacities. Stem cells are often named for the part of that organism they build and/or replenish: neural stem cells, hematopoietic stem cells, epithelial stem cells, muscle stem cells, and so forth. In sum, multicellular organisms are not only sources of stem cells, which thereby partly determine the latter's reproductive and developmental potential, but also implicated in the realization of that potential. The concept of a stem cell is "flanked by" multicellular organisms, as source, context, and developmental product.

A Philosophical Model

Let us draw these different strands together. The original Stammzelle concept was associated with cell and phylogenetic lineages; although the latter is no longer applicable, the former is central to the stem cell concept today. In scientific practice, "the stem cell" is a conceptually untidy category, encom- passing many varieties with no clear system of classification. Introductory and public-facing definitions offer more guidance, converging on two cell lineage- producing abilities: self-renewal and differentiation. All the disparate varieties of stem cells are capable of both, to some degree or extent. This schematic

---

[43] In "Rediscovering the Hydra," Cooper (2003) discusses multiple ways that insights from stem cell research overturn Weismannian ideas about development and regeneration.

characterization balances the general functional definition with stem cell diversity: two essential abilities, realization of which is highly variable. In addition, multicellular organisms are implicated in the stem cell concept; stem cells both come from and produce parts of multicellular organisms.

These ideas are integrated into an abstract philosophical model of the stem cell concept: a simplified conceptual structure that captures the idea of a stem cell, as the latter figures in scientific practice. Though grounded on the definitions discussed previously, this philosophical account is more explicit, precise, and abstract. Here some disciplinary differences should be noted. The *idea* that the model is meant to capture is scientific. But the model itself has no counterpart in stem cell biology. Some areas of science construct and work with abstract models, and philosophers of science have gained much insight from examining the practices of model-construction and use in theoretical physics, population genetics, systems biology, and other fields. Modeling practices in stem cell biology are quite different, however. Stem cell researchers make little use of abstraction and idealization in model-building.[44] A step toward abstract modeling of the stem cell concept is a step away from the practices of stem cell research. The philosophical model proposed here is for philosophers (and other nonexperts), not for stem cell scientists. It connects to the latter's practices *indirectly*, by further explicating the definitions offered to outsiders in terms of the cell-organism relation and lineage structures generated via stem cells' essential abilities. So my philosophical model departs from scientific practice in at least two ways: it is constructed by methods not used in stem cell research, and it is grounded on definitions not used by scientists in their own work.

What is the point of such a model? Why not clarify the stem cell concept by examining stem cell research practices directly?[45] The answer reveals my own commitments about how to do philosophy of science in practice. Stem cell research is variegated, technologically sophisticated, jargon-laden, and thoroughly disunified. In virtue of these features, it is largely impenetrable to nonexperts (like many areas of science today). To engage it in the way scientists do requires learning a lot of biology: molecular, cellular, developmental, structural/biochemical – and other fields as well, such as cancer biology, immunology, hematology, and (increasingly) bioengineering. Some philosophers of science do have this expertise, but not many. Because there is no systematic classification of all stem cell varieties, nonexperts have little hope of engaging with stem cell science, short of gaining hard-won expertise themselves. An important role for philosophy of science, in my view, is to articulate the

---

[44] The central models in stem cell research are *concrete*: cultured cell lines, model organs and embryos, and so forth. I discuss these in Section 3.

[45] Thanks to an anonymous reviewer for raising this question.

conceptual structure and methods of current science – especially for fields with high social significance and impact – for wider audiences, mediating between these epistemic communities. So the model I propose is a construct that (I think) reveals that conceptual structure, in a way that makes features of stem cell research intelligible for outsiders. This is why my model builds on the public-facing definition, used to communicate the basic idea of a stem-cell to non-experts. I have elaborated on that simple definition in several ways, bringing in ideas of reproductive and developmental cell lineages, and the cell-organism relation – ideas implicit in the stem cell concept since its inception. In this way, the model sketches the conceptual anatomy of stem cell research, creating a point of entry into the field for nonexperts. I first present the model, then discuss its role in mediating between stem cell science and philosophy.[46]

A stem cell:

(1) is derived from a context within organismal source S, where S is characterized by species and developmental stage;

(2) is the origin of an actual or potential reproductive cell lineage, the structure of which is determined by n cell division-events preserving sameness with respect to characters C;

(3) is the origin of an actual or potential developmental hierarchy, the structure of which is determined by (i) the number of stages; i.e., depth of developmental hierarchy, (ii) the number of developmental end points (i.e., a cell's developmental potential), and (iii) the arrangement of branch-points linking successive stages; i.e., developmental pathways; and

(4) can contribute to an organismal product P.

Condition (1) represents key features of the organismal source, which largely determine the specifics of a stem cell's capacities for self-renewal and differentiation. All stem cells originate in some organismal context: a particular species, developmental stage, and location from which stem cells are derived. Conditions (2) and (3) specify the two defining stem cell capacities in terms of cell lineages. As shown previously, both self-renewal and differentiation produce cell lineages; they are lineage-producing processes. A lineage, recall, is a complex biological entity (or process) composed of lower-level entities linked by ancestor-descendant relations, which takes the form of a linear sequence or branching tree. To be a stem cell is to be the origin of both kinds of cell lineage, either actually or potentially. This follows from the meanings of "self-renewal" and "differentiation," discussed

---

[46] This model builds on my previous work (Fagan 2013a, 2013b, 2016b, 2017, 2019). The earliest version (2013a, 2013b) generalized the working definition in terms of "asymmetric cell division" and did not include features of the organism; for other contrasts with later treatments, see Fagan (2019).

previously. To be a stem cell of a particular variety – the way the concept is used in scientific practice – is to be the origin of lineages with the particular structures characteristic of that variety (i.e., specific degrees/extent of self-renewal and differentiation potential). Condition (2) is a schema for the structure of a cell lineage generated by self-renewal. Self-renewal generates a lineage of stem cells through repeated rounds of cell division that preserve distinguishing characters (C) of that stem cell variety. Such characters may include the rate of cell division, cell size and shape, the relative size of the nucleus and cytoplasm, and the expression level of particular genes – whatever phenotype is assigned to that variety of stem cell. The extent of a stem cell's self-renewal capacity fixes the life-span of a stem cell lineage. This extent (n) varies across different types of stem cell, from a few weeks to (apparently) no upper limit ("immortal" stem cell lines).[47] A lineage generated by cell division is obviously thickly branching; its form is determined by the number of cell division events.

The idea of a developmental lineage is more abstract. "Ancestors" are cell types that appear earlier in development than their "descendants." The form of a developmental lineage-tree need not track bifurcating cell divisions. Differentiation produces a developmental hierarchy of cells, organized into pathways of phenotypically distinct stages. Many different tree-topologies are possible. Any tree-topology can be characterized in terms of a number of stages, of termini, and the arrangement of branch-points. So the possible structures are captured by variables specifying the structural features of any tree-topology with a single origin:

(i) d, number of stages (i.e., depth of hierarchy);
(ii) p, number of end points or termini; and
(iii) a, arrangement of branch-points linking the origin to end points.

Condition (3) is a schema for the structure of any developmental lineage initiated by a stem cell. Assignment of values to variables d, p, and a specifies the structure of the developmental lineage to which a stem cell can give rise; these values pick out a tree-topology from the space of possible lineage tree models. Each model within this space has a particular number and arrangement of developmental stages, termini, and branch points – a particular structure, or tree-topology. The arrangement of branch-points indicates the pattern of distinct developmental pathways initiated by a stem cell. The number of termini corresponds to a stem cell's developmental potential. The number of distinct developmental stages in each pathway determines the depth of the hierarchy.

---

[47] When self-renewal ceases, the stem cell population will steadily deplete through differentiation and cell death, followed by the descendant populations.

Condition (4), for now, is a placeholder. I have argued that a stem cell's developmental potential involves the whole organism as well as cells. Condition (3) is a schema for a cell lineage only. There is more to say about stem cells' developmental potential, but to do so requires more detail about stem cell experiments. That is a topic for the next section. The meaning of Condition (4) will be fleshed out then. For now, I will sum up the insights of this philosophical model of the stem cell concept. To be a stem cell is to be the origin of a reproductive cell lineage of n cell generations, and a developmental cell lineage of some form or other, derived from some location within an organismal source of particular species and developmental stage, and capable of contributing to some organismal product (Figure 6). This account reflects the main ideas clarified previously: self-renewal and differentiation are central, cell-organism relations are (somewhat) clarified, and the wide variety of stem cells is accommodated by the multiple variables of Conditions (1–3). Any stem cell (or population, or variety) can be characterized in terms of its organismal source, lineage life-span set by the extent of self-renewal, and the structure of the developmental cell lineage it initiates (Table 1). Different varieties of stem cells occupy different regions in a space of possible lineage tree models, which includes all possible combinations of values of variables in Conditions (1–4).

One limitation of lineage models is that cells and their descendants are represented as isolated from their environment (i.e., the "cell niche" emphasized in scientific accounts of stem cells). This suggests that stem cells are intrinsically endowed with lineage-generating capacities, autonomous and independent of environmental factors. But this is not so (as discussed in the next section). A stem cell can generate reproductive and developmental lineages only under specific environmental conditions. Stem cells' generative abilities are context-dependent, not absolute. Indeed, much stem cell research can be summarized as aiming to discover the precise environmental

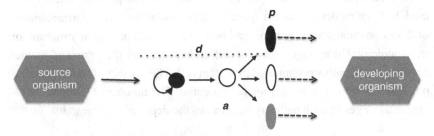

Figure 6 Schema of abstract stem cell model

Table 1 Abstract stem cell model covering several major stem cell varieties.

Type	S	C	n	D
ESC	5d embryo ICM (human)	cell size, cell shape, gene expression, karyotype, telomerase activity, alk-phos, cell surface molecules	$\geq$50 divs	all cells descended from three germ layers; post-embryonic developmental pathways
HSC	BM, cord, peripheral blood	cell size, density, light scatter, surface molecules, cell cycle status	>6 months	blood and immune cell lineages
NSC	basal lamina of ventricular zone	cell morphology, surface markers, gene expression, cytokine response	months to years	cell lineage terminating in neurons, astrocytes, and oligodendrocytes
iPSC	various (relatively mature cells)	colony shape, cell size, cell shape, nucleus/cytoplasm ratio, cell surface molecules, activity and expression of specific proteins, gene expression (specific and global), histone modifications at key locations	$\geq$50 divs	all cells descended from three germ layers; post-embryonic developmental pathways

Table 1 (cont.)

Type	S	C	n	D
GSC	5-9wk gonadal ridge	colony shape, alk-phos, surface expression (SSEA-1, SSEA-3, SSEA-4, TRA-1–60, TRA-1–81)	20-25 wks	all cells descended from three germ layers
EC	teratocarcinoma (129)	cell shape, morphology, production of embryoid bodies, surface molecules, enzymes	unlimited	some cells from each of three germ layers1

conditions to reliably generate some desirable product from stem cells.[48] Stem cells and their lineal descendants are no more separate from their environments than the populations and species depicted in phylogenetic trees. The flanking organismal conditions (1) and (4) do not fully capture this context-dependence. In this respect, the stem cell concept has not yet been brought down to earth so to speak (i.e., grounded in experiments). The philosophical model does, however, offer a framework for engaging those experiments and their results. As noted, the idea of a stem cell in actual . research practice is highly disunified, technical, and detailed – dispersed into a welter of varieties, each with an enormously detailed (and rapidly changing) scientific characterization. Appropriately, then, my philosophical model does not provide a set of necessary and sufficient conditions for being a stem cell, as that concept figures in scientific practice. There is no such set of conditions – at least, none grounded in stem cell science. (And what other grounds could reasonably offer such conditions?)

The abstract model does not pick out all and only stem cells that are known to science or yet to be discovered. At this time, there is no such unified account of stem cells – and perhaps there never will be.

In another respect, however, my philosophical model of the stem cell concept fits elegantly with experimental methods of stem cell research. The abstract model includes variables that specify structural and substantive features of a stem cell's organismal source, reproductive and developmental cell lineages, and organismal products (the last a "black box" for now). Although Conditions (1–4) are not individually necessary and jointly sufficient for "stemness," the model provides a general abstract framework for conceptualizing all the different varieties of stem cells (Table 1). The variables can take different values, accommodating all the different varieties of stem cells. In scientific practice, the values of these variables are specified by the materials and methods of experiments aimed at identifying and characterizing a particular kind of stem cell. The philosophical model is a general schema to which different experiments can be fit by specifying values of some or all variables in the model. In this way, the model provides a starting point for more detailed characterizations of particular varieties of stem cell: adult, embryonic, pluripotent, induced, neural, epiblast, hematopoietic (blood-forming), and more. Philosophers can thus use the abstract model to engage the field, without immediately becoming enmired in its daunting technical terminology, rapid pace of change, and multiple, overlapping, cross-cutting meanings of "stem cell." Many strands of stem cell research can be understood as pursuing different specifications of the same abstract,

---

[48] For example, hair follicle and intestinal stem cell regulation via the niche (Mesa et al. 2015).

schematic model. This framing makes possible a kind of overview, not easily obtained from reading the scientific literature on stem cells. That's the purpose of the philosophical definition – to aid philosophers, and other nonexperts, in engaging stem cell science. Doing so promises to be fruitful for a number of ongoing philosophical debates. I conclude by indicating some of these topics for future work.

Further Questions

This section has examined the concept of a stem cell, tracing the term's historical roots to current stem cell research. The latter is very diverse, comprised of studies of many different varieties of stem cells. Inconveniently for philosophers, there is no clear classification system for stem cells. Instead, we are presented with an array of cross-cutting and overlapping categories: adult, embryonic, fetal, induced, hematopoietic, mesenchymal, germline, epiblast . . . on and on. To make headway, I have proposed a philosophical analysis of the stem cell concept, in the form of an abstract model representing key features of stem cells as variables. This clarifies the focal concept of stem cell research and offers a window into the field through which philosophers can become acquainted with this science. To conclude this section, I discuss other significant philosophical questions posed by stem cell research and the very idea of a stem cell. Answers are a task for future work.

Biological Individuality

Most philosophical accounts of biological individuality prioritize evolutionary theory as the ground of criteria for determining what counts as a biological individual. Life sciences offer many examples of hard cases against which philosophers' accounts are sharpened: biofilms and other microbial phenomena, clonal organisms, symbioses, "superorganisms" integrated by sociality, and more. But life sciences offer more than examples for philosophers: conceptual resources can be extracted from fields not connected to evolutionary theory. Pradeu (2012) draws on immunology's rich theoretical debate on the self/non-self distinction to propose an innovative "physiological" account of biological individuality. Stem cell research offers promising conceptual resources along similar lines. My own work in this area is only a small start on the many avenues that could be pursued.[49] How does biological individuality for cells relate (if it does) to biological individuality for organisms? How (if at all) do cells comprising the body of a multicellular organism differ from cells comprising the

---

[49] For example, Fagan (2016a, 2018).

holobiont? How does the process of development differ (if it does) for free-living single-cell organisms, cells of a multicellular organism, and microbial symbionts in a holobiont? In what sense are cell lineages individuals, and how does this bear on our ideas about biological development? Are cancers individuals in any meaningful sense of the term? If so, how does "malignant individuality" differ from that of an organism? These questions and more arise from reflection on the basic concepts of stem cell biology.

Classification and (Un)natural Kinds

Natural kinds are a perennial topic of interest for philosophers of science – and here the kind "stem cell" poses an interesting challenge. "Cell" is a good candidate for a natural kind (see section 1) – but what about stem cells? I have noted several times the lack of any systematic classification of these entities. Stem cell research is marked by a profusion of varieties, but no single set of categories to distinguish them. Instead, there are many different, partly overlapping and cross-cutting, kinds of stem cell. A number of these are human-made, rather than naturally occurring. Julia Bursten (2019) suggests that this pluralistic, disorderly character may be a general feature of scientific kinds in experiment-driven fields like stem cell biology and nanoscience – perhaps especially associated with practices of "destructive experimentation" (see the next section). Parallels with classification in engineering fields, such as nanoscience, where novel entities and processes are synthesized, could be a fruitful way to investigate classification in stem cell biology. Relatedly, Catherine Kendig (2016) suggests that philosophers focus on "kinding" practices: ways that scientists individuate and classify objects of inquiry. Along these lines, scientific proposals for stem cell nomenclature intended for biobanks and data repositories offer significant philosophical resources. For example, Kurtz et al (2018) propose a nomenclature system for human pluripotent stem cell lines incorporating five elements: (1) the laboratory/institution, (2) stem cell variety, (3) human donor, (4) specific cell line (of multiple clones from a donor), and (5) genetic modifications generating a sub-line from a given clone.

Stem cell–initiated lineages also offer prospects for philosophical enquiries into classification in life science. Stem cells, by definition, are the starting points of cell lineages. (So I argue, anyway.) The same features by which these peculiar entities are classified as varieties of stem cells determine a lineage structure than can, in turn, be used to classify cells within that lineage. Phylogenetic lineages are the prevailing, although not the only, mode of classification for organisms and species. Classification of the major cell types, as noted previously, does not traditionally make use of lineages — but classification within those types often does (e.g., neurons, blood, and immune cells). Stem cells' distinguishing developmental capacity, the potential to differentiate into a range of other cell types, might itself be deployed as a classificatory principle for cells of the associated lineage. In this way, stem cell classification might feed back to inform classification of more traditional cell types, and perhaps of developmental modes pertaining to tissue, organ, and organismal levels of biological organization.

Metaphysical Issues

The idea of *potential* is central to stem cells – but what is its metaphysical import? How should the concept of potential be understood? The interplay of potential and actual is obviously central to the nature of stem cells and their biological role. To be a stem cell is to be the origin of a cell lineage, either actually or potentially. Haeckel's early statements make the point evocatively: "The ovum stands potentially for the entire organism—in other words, it has the faculty of building up out of itself the whole multicellular body … it unites all their powers in itself, though only potentially or in the germ" (Haeckel 1905, 103–104).

Aristotelian ideas could be useful in clarifying stem cell potential and its relation to actual processes.[50] A related debate is whether stem cells are entities or processes (Dupré and Nicholson 2018). Although cells would seem to be prototypical entities, there is a case to be made that stem cells are processes. Stem cells are defined by processes of cell reproduction and development that they initiate. Both processes are transformative and future-oriented. The very idea of a stem cell invokes the processes that generate a cell lineage with a particular structure. Stem cells' extreme context-dependence also undercuts a substantivist view. Decades of experimentation have shown that being a stem cell is exquisitely sensitive to features of the local micro-environment, both in vivo and in vitro (see Section 3). A substantivist interpretation of the model is liable to elide this context-dependence. This in turn tends to encourage a mistake – thinking of stem cells as a stable, fixed kind of entity, rather than relational, context-dependent, and individuated relative to conditions imposed by experimental methods. A process interpretation, in which the stem cell actively generates a lineage with a more or less precisely specified tree structure, is much more consonant with the actual experimental practices by which stem cells are individuated (in all their varieties). In addition, the cell-organism relation may be better captured by a process view than a substantivist one. However, these considerations do not rule out

---

[50] Anne Peterson, personal communication.

a substantivist view (e.g., that self-renewing stem cells are conceived as a stable cell type, with the causal power of producing other cell types). Stem cell biology offers interesting prospects for this and other "metaphysics of science" projects.

Conclusion

The previous section indicates the range of philosophical topics that remain to be explored through the lens of stem cell research. Each affords prospects for engaging the large existing literature on stem cell ethics and policy issues as well. This brief treatment can do no more than gesture at these projects. This section has focused more narrowly on the idea of a stem cell, adding philosophical abstraction to definitions used in scientific practice, past and present. I have reviewed the original meaning of "Stammzelle" and surveyed several present-day working definitions offered to practicing scientists, novices, and the public. Building on the latter, I proposed a philosophical definition, which takes the form of a simple, abstract model. The model clarifies the general concept of a stem cell, for a philosophical audience. Stem cells are *essentially generative*: they are cells defined by their ability to produce other cells, both like and unlike themselves. This generative, future-oriented aspect is conceptualized as stem cells' *potential* or *potency*. A stem cell's reproductive potential is realized in self-renewal; i.e., cycles of cell division that produce a clonal population of the original stem cell. A stem cell's developmental potential is realized in a process of differentiation, which generates a developmental hierarchy. Both processes take the form of a lineage. The final result: a stem cell is the origin of reproductive and developmental cell lineages, derived from and producing (parts of) a multicellular organism. The next task is to bring this concept down to earth, to biological reality. How do we find things that fit the idea of a stem cell? More succinctly, how do we gain knowledge of stem cells? The next section examines this question.

3 Finding Stem Cells

Our only access to stem cells is through *experiment*. Experiment is something of a cipher in philosophy of science. Traditionally, it is relegated to a supporting role: a source of evidence, along with observations, for hypotheses and theories. With the rise of causal philosophies of science, experiment is increasingly conceived as the means of discovering causal relations in datasets. Historically informed studies of experiment are richer and more nuanced.[51] Rather than treat experiments solely in relation to other

---

[51] For example, Hacking (1983), Steinle (2002), and Chang (2004).

philosophical topics, such studies examine experiments on their own terms, as part of scientific practice. That is the approach taken here. Many, many different kinds of experiment figure in stem cell biology: clinical trials, ex vivo cell culture, antibody staining, functional genomics and other bioinformatic techniques, knock-out/-down/-in genetic manipulation, cell reprogramming, machine learning, single-cell lineage tracing, tissue engineering, fluorescence-activated cell sorting, computer simulation, gene editing, and more. The inventory is long and open-ended, as new technologies bring forth continual additions. On the other hand, some formerly influential experimental methods have passed out of use, superseded by technological innovations. A comprehensive account of experiments in stem cell research would fill a much longer volume than this one (and would likely be outdated before publication). Instead, this section surveys three kinds of experiment aimed at identifying and characterizing stem cells – that is, experiments that ground the abstract idea of a stem cell in biological reality. This connects closely with the key results of Section 2 and provides an overview of what I take to be the conceptual center of stem cell research.

Experiment without Theory

Traditionally, the central task for philosophy of science is the clarification of physical theories and their evidential basis. A *theory*, in philosophy of science, is an abstract structure that can be represented as a set of statements, system of equations, and/or simple diagrammatic pattern. Philosophers might expect that proliferation of experimental methods would be necessarily accompanied by an efflorescence of theories, in this traditional sense. It is widely though not universally assumed among philosophers of science that our best scientific knowledge takes the form of abstract and concise structures, typically mathematical. Stem cell biology confounds these expectations. Knowledge of stem cells and related phenomena does not take the form of general theoretical statements, equations, or patterns. The field has no obvious counterpart to theories in physics or neo-Darwinian evolutionary biology; its methods and results bear little resemblance to those sciences, the favorites of philosophers. Rather than theory, stem cell research is driven by experiment – motivated and guided by available technology and hoped-for applications. As noted in the previous section, stem cell researchers tend to eschew abstract models and use general rules (e.g., the correlation between the cell and organismal potential) only as highly defeasible background assumptions. Knowledge in stem cell research takes the form of experimental innovations: new ways of finding, growing, or making stem cells. This is not to say, of course, that stem cell biology does not use concepts or lacks conceptual structure, nor that inference and reasoning play no role in the field. It is just that those concepts, inferences, and reasoning are focused on experiments rather than the more abstract representations philosophers are drawn to. Our knowledge of stem cells is very closely tied to its experimental ground. The goal of this section is to articulate that ground: the main experimental methods by which we gain knowledge of stem cells, their distinctive features, and limitations. As in section 2, I balance engagement with the science on its own (experiment-driven) terms with the philosophical task of clarification. Experiment-driven fields like stem cell research do not simply lack theories in the traditional sense. There is a positive philosophical insight to be gained by explicating their methodological and epistemic structure on its own terms. The last part of this section discusses several such insights, which connect with recent philosophical work on models and social epistemology of science.

Find, Grow, or Make

Although myriad experiments comprise stem cell research, a few stand out as exemplars. These are recognized as groundbreaking for the field, adopted by many laboratories after their initial innovation. At least three experimental exemplars have profoundly influenced stem cell science to date, giving rise to widely applied strategies of *finding*, *growing*, or *making* stem cells. Methods for finding stem cells descend from the first quantitative blood stem cell assay (Till and McCulloch 1961). The exemplar for growing stem cells is the method for culturing human embryonic stem cells in vitro (Thomson et al 1998). The original exemplar for making stem cells is the method of "reprogramming" to induce pluripotency (Takahashi and Yamanaka 2006). I survey each, then extract some general points.

Finding Blood Stem Cells

Throughout an animal's life, blood stem cells (a.k.a. hematopoietic stem cells, or HSC) replenish the short-lived cells of blood and the immune system. HSC are found (in post-natal mammals) in bone marrow, circulating blood, and umbilical cord blood. They were the first variety of stem cells to be characterized, the first used in routine clinical practice, and remain the best understood of all stem cell varieties. Although the idea of a blood stem cell in bone marrow dates back to the 1890s, the first experimental assay for HSC emerged more than fifty years later. This experimental exemplar was an offshoot of post-World War II radiation research, specifically the discovery that lethally irradiated mice can be "rescued" by injecting bone marrow cells from a donor mouse.[52] Bone marrow transplantation prevents the painful death that would otherwise ensue from high doses of gamma-radiation. James Till and Ernest McCulloch, researchers at the Ontario Cancer Institute (Toronto), noticed that spleens of transplanted mice were lumpy with "nodules," unlike smooth normal organs.[53] The spleen filters blood, so it was to be expected that injected bone marrow cells would pass through it. Apparently, a few donor cells lodged in the spleen and gave rise to nodules there. The nodules were large – about a million cells each – yet chromosome markers showed that each was a "clone" generated by division from a single cell.[54] When dissected, nodules were found to contain all types of blood and immune cells known at the time. It followed that this "colony-forming cell" had the developmental potential to produce all blood and immune cells. Self-renewal was inferred as well, bolstered by the observation that cells from spleen colonies can rescue irradiated mice and produce new spleen colonies in their turn (i.e., colony-forming ability can be propagated for weeks or months). Therefore, the single cell initiating a spleen nodule is capable of both self-renewal and differentiation – a good candidate for the long-hypothesized-but-never-observed HSC.

Till, McCulloch, and their colleagues invented a quantitative assay to measure these stem cells (whose presence was only inferred from studies of spleen nodules). They found an approximately linear relationship between the number of bone marrow cells injected and the number of spleen colonies formed: about 10^4 cells per colony. So, which of those 10^4 cells was the "colony-forming unit"? Answering this question amounted to panning for regenerative gold in the bloodstream (with the spleen as a filter). Inconveniently, stem cells do not reveal themselves like shining particles of gold or even other cell types that display distinctive traits under a microscope. The only way to tell if one has isolated a bone marrow cell with stem cell capacities is to realize those capacities. The experiment designed to identify and characterize HSC was the template for nearly all other efforts to find stem cells in an organism's body.[55] The

---

[52] After the invention and use of atomic bombs during World War II, there was considerable scientific interest in learning how to counteract lethal effects of radiation. The interface of radiation research, immunology, and hematology in the mid-twentieth century was a crucible for studies of HSC (see Kraft 2009, Fagan 2011, Sornberger 2011).

[53] James Till and Ernest McCulloch were a collaborative team of statistician and radiation biologist, respectively.

[54] This is the original scientific meaning of the term "clone": a colony of genetically identical cells related by reproduction, with connotations of genetic identity and simple genetic causation (both now scientifically dubious notions).

[55] More recent experiments use single-cell sequencing and lineage tracing to identify HSC and other stem cells, retrospectively. For now, classic HSC methods are still influential in stem cell biology, not least in setting epistemic standards for the field.

method has four steps, which when performed in any particular experiment specify the four conditions of the abstract model (Section 2). First, bone marrow cells are extracted from a healthy donor mouse. Next, extracted cells are sorted into subpopulations, each to be tested separately. Third, some number of cells from each subpopulation is injected into a lethally irradiated mouse. After a time interval (10–11 days in the original experiments), surviving mice are killed, dissected, and the number of spleen colonies counted. This final step measures each subpopulation's colony-forming ability. The goal is to confine that ability to just one subpopulation of bone marrow cells. But in the first efforts to identify HSC, there was no set of distinguishing characteristics to use in creating subpopulations and no principle of selection for sorting bone marrow cells in a way that can reveal HSC. Colony-forming cells were known only through their stem cell capacities; their cell traits would be discovered through these very experiments, if at all. To sort bone marrow cells into subpopulations, researchers used a combination of educated guesswork and trial-and-error. The standard for success was "enrichment" of HSC capacities, relative to unsorted bone marrow and other sorting methods.

The search for HSC was a worldwide community effort, not a task for a single laboratory. No one has direct insight into the distinctive traits of blood stem cells, or any other kind of stem cell. There is no abstract predictive system one can use to pinpoint those traits, bypassing the hard concrete work of experimentation. HSC researchers just did the work, in a distributed and rather uneven way.[56] Over the next three decades, Till and McCulloch's experimental method – their quantitative assay plus the cell sorting step – dispersed into many different laboratories. Research teams across the world tried different combinations of cell traits, ran those subpopulations through the spleen colony assay, and used statistical arguments to show (or try to) that the numbers of resulting spleen colonies were consistent with a pure starting population of HSC. Nearly any trait that could be used to physically sort cells into distinct subpopulations without harming them was tried: size, density, cell cycle status (i.e., dividing or nondividing), and presence or absence of specific cell surface molecules (e.g., CD34, c-kit, Sca-1). The hoped-for result was a set of traits associated with all and only long-term regenerative potential for the entire blood and immune system. Such a list of traits would amount to an HSC "profile" that could be applied to physically pick out just those cells from an organism's body.[57] This

[56] The community structure of experimental research is a ripe topic for social epistemology modeling: how might such communities be organized and exchange information for more efficient problem-solving of this kind?

[57] An interesting methodological contrast with analytic philosophy: an HSC profile, or any other stem cell variety's profile, looks a bit like necessary and sufficient conditions, the sought-after

was accomplished in the 1980s – but not in a way that yielded a single, unambiguous HSC characterization. There is no such characterization even now; no single cellular entity, "the HSC" with a fixed set of traits. Instead, there are diverse ways of identifying and characterizing HSC, corresponding to different sorting and testing criteria. The label "HSC" is ambiguous.[58]

This outcome may seem puzzling. Why did the far-flung trial and error experiments to isolate HSC not quickly converge on a single cell population, "the real HSC"? Part of the answer is technological change. New innovations for separating bone marrow cells into subpopulations were incorporated into the original method, while new in vitro cell culture assays eventually replaced the original spleen nodule "read-out."[59] More recently, gene expression profiles have come to dominate the list of characters associated with HSC capacities. New possibilities for cell sorting and testing developmental potential continually refresh the basic method of finding HSC. Another part of the answer is that our grasp on stem cells in organisms is very indirect. Ingenious, convoluted experimental design is required to get around stem cells' in vivo elusiveness.[60] But with this convoluted design comes uncertainty. That uncertainty contributed to prolonged debate about how to identify HSC and whether any particular experiment had done so. (I return to this issue shortly.) Evidently, finding stem cells within an organism's body is a tricky exercise. A more direct approach is to grow cells with those capacities outside the body: in vitro. This brings us to the second exemplar.

Growing Embryonic Stem Cells

An alternative to seeking elusive stem cells within an organism is to extract an organism's cells and grow them in artificial cell culture to produce an in vitro stem cell line. Trial and error is again needed because we do not know in advance the right environmental conditions to grow stem cells of any particular

---

form of conceptual analysis. It is tempting to conflate the two – but that would be mistaken. Stem cell researchers are not trying to isolate "all and only" HSC in bone marrow with analytic precision. It does not matter if they miss some, nor if a few other cells sneak in. What they want is a way to isolate enough HSC to reliably get effects they are interested in. Varying interests is one reason for the persistent ambiguity of "HSC" (see the main text for more).

[58] Crisan and Dzierzak survey "the many faces of HSC heterogeneity" in a recent review (2016).

[59] The most important technological innovations for cell labeling and sorting were monoclonal antibodies and fluorescence-activated cell sorting (FACS), both developed in the 1970s. Monoclonal antibodies distinguish morphologically similar cells by binding surface proteins with exquisite specificity, without compromising other cell functions. FACS separates living cells one-by-one into distinct subpopulations on the basis of antibody-labeled surface molecule expression and other physical properties (Keating and Cambrosio 2003).

[60] Some varieties of stem cell, for example mouse intestinal crypt stem cells, are more straightforwardly identified by anatomical location and expression of certain genes. This is an antecedent of single-cell lineage tracing methods, a new experimental approach that can only be briefly sketched in this short treatment.

variety. (Mammalian cells – and human cells especially – are notoriously finicky about their growing conditions.) On the other hand, this experimental strategy has a century-long tradition of cell culture techniques to draw on. Artificial cell cultures standardly consist of a transparent dish or flask of liquid media enriched with nutrients, kept at a constant (warm) temperature (Landecker 2007). The nutrient-rich liquid mimics conditions inside the body – not exactly, but a close enough approximation for cultured cells to divide and multiply. Often, the culture environment is designed to encourage self-renewal: clonal reproduction of the same cell type. Many, many different kinds of cells can be cultured in vitro: bacterial colonies, mouse cells secreting specific antibodies, synchronized beating hearts in plastic dishes, and incalculably more. The technique is very widespread in life and medical science (although largely ignored by philosophers). Different kinds of cells thrive under different culture conditions. What knowledge do we get from cell and tissue cultures? The simplest answer is that it is a form of scientific modeling: cells are removed from their naturally occurring context and placed in an artificial environment where their structure and function can be studied more easily. Lost, of course, are any interactions with the natural cell niche that are not approximated by the artificial situation. Because much about naturally occurring cell niches is unknown (especially in mammals, even more so in humans), we cannot reliably reconstruct all the relevant interactions. So cell culture involves a tradeoff: cells are accessible to observation and experiment, at the cost of losing (largely unknown) contextual factors.

This epistemic tradeoff fits well with an influential philosophical view about models in science. Very briefly: scientific models are simplified, idealized constructs, relative to some target of interest, that support inferences about that target ("surrogate reasoning"[61]). To underwrite those inferences, the model-target relation must satisfy certain criteria. Philosophers continue to debate those criteria, but in many cases, something like similarity or analogy between model and target seems required. In addition, there is a strong consensus that what is required for a model to be successful – that is, exactly how and in what ways it should relate to its target – depends on that model's purpose (or, more accurately, on *model-users'* purposes).[62] Models play a lot of roles in science, and their standards for success are accordingly variable. Use of any model can be analyzed as a four-place relation: "Agents (1) intend; (2) to use model, M; (3) to represent a part of the world W; (4) for purposes, P. So agents

---

[61] See Godfrey-Smith (2006) and Weisberg (2013).

[62] Model-users are epistemic agents, with purposes, values, and interests largely shaped by their communities. Studies of scientific models thus connect to social epistemology of science, although philosophers have been slow to exploit this link.

specify which similarities are intended and for what purpose" (Giere 2010, 274). Applying this general account to the case at issue: cultured cells are concrete models of cells in their natural habitat, with myriad different uses. An important use for cultured *stem cells* is to learn about the developmental capacities and regenerative products of stem cells in the body – elusive entities that, as we have seen, resist unambiguous characterization. Similarities between in vivo and in vitro environments, and therefore between cells in those respect-ive niches, are vital for cultured stem cells to achieve that modeling purpose. But, once removed from the body, they also become objects of scientific interest in their own right. Many stem cell researchers aim to characterize developmen-tal capacities and regenerative products of cultured stem cells directly, inde-pendent of their modeling role.[63] This is typical of scientific modeling: models tend to become research foci somewhat independent of their targets.[64] Stem cell researchers gain knowledge both of cultured stem cells as such (direct objects of study) and of mammalian cell, tissue, and organ development (indirect targets of model-based research).

Among the various cultured stem cell models, human embryonic stem cells (hESC) are the central exemplar. Although not the first stem cells cultured, hESC have profoundly impacted and shaped the field of stem cell research.[65] Their origination in 1998 reoriented the field of stem cell research around embryos, not only spurring bioethical debate but also instituting stem cell research as a distinct scientific community rather than a branch of hematology (and a smattering of other biomedical fields). hESC were first successfully grown by a multi-national group led by James Thomson at the University of Wisconsin (1998). Their method built on a background of experimental success in culturing other varieties of stem cell, notably those grown from embryos of mice and non-human primates. Conditions for growing hESC were discovered over several decades, by a process of worldwide community trial-and-error and rapid uptake of successful culture recipes. So this exemplar is not itself a pioneering method, but a striking application and modification of existing culture methods. Thomson and colleagues obtained early human embryos (about five days, or blastocyst stage) from IVF clinic patient donors.[66] Blastocysts have an outer layer (trophectoderm), which can give rise to extra-embryonic tissues, and an inner cell mass (ICM), which can give

---

[63] Independent at least in the first instance (as discussed later).

[64] Medical applications, however, bring the focus back to the original targets – notably, cells developing and functioning in human bodies (see Section 4).

[65] The term "embryonic stem cell" was coined by Gail Martin, one of the first researchers to grow mouse ESC in culture (1981).

[66] These are, obviously, in vitro human embryos – the entire method takes place in artificial culture.

rise to embryonic tissues. From in vitro early human embryos, the method proceeds as follows. First, a chunk of ICM is removed and placed atop a "feeder layer" of cultured mammalian cells in a petri dish. In this new artificial environment, bathed in secreted growth factors from the feeder layer, some extracted cells divide rapidly, producing colonies after one or two weeks. (The original feeders were mouse cells, then human cells, now replaceable by a defined culture medium with a few chemical additives.) Colonies are picked off the plates, dissociated, and a few replated, so new colonies appear in turn. Thomson and colleagues looked carefully at the traits of cells and colonies, selecting those that exhibited rapid division in culture, lack of specialized cell structures, flat round shape, large nuclei surrounded by correlatively thin cytoplasm, prominent nucleoli, high telomerase activity, and high expression of a few specific genes highly expressed in early embryos but not in later developmental stages. This yielded a collection of "immortal" self-renewing cell lines with the selected traits.[67]

Self-renewal is built into this method: cells divide to form homogeneous colonies on "generations" of cell culture plates.[68] hESC exhibit unlimited self-renewal, by experimental design. Differentiation potential, however, needs to be tested, to show that hESC have embryo-like developmental capacities. ICM cells are pluripotent: capable of giving rise to all the cell types of an adult human body. But in embryos, mammalian cells quickly lose this capacity, becoming more restricted in differentiation potential as organismal development proceeds (see Section 2). Thomson and colleagues hoped to grow human cells that retain pluripotency in the long-term, in effect freezing them at an early embryonic developmental stage. This would offer an unprecedented view of early human embryonic development – albeit indirectly, through a concrete model. In fact, it proved more challenging to prevent cultured ESC from differentiating than not – these cells "want to differentiate."[69] To demonstrate pluripotency, samples

---

[67] It is not usually recognized in bioethical debates, but what happens to an early embryo in these experiments is not "killing" in any familiar sense of the term. The embryo does not develop as it would in a natural context, but it is not discarded or terminated. Quite the contrary: ICM cells removed from the early embryo are propagated indefinitely in cell culture. These embryonic cells remain alive, giving rise to an "immortal" cell lineage with the prospect of contributing to incalculably many human organisms at a variety of developmental stages. It is another form of life, but the embryonic cells do not end as other remnants of IVF procedures do. That is the whole point of growing stem cell lines. See Skloot (2010) and Landecker (2007) for thoughtful discussions of the life of cells and of human persons.

[68] Note, too, how Conditions (1) and (2) of the abstract model are specified by the procedure.

[69] Cultured pluripotent stem cells tend to differentiate spontaneously if left to themselves, often in uncontrolled and unanticipated ways. This occurs even in artificial culture that encourages self-renewal, particularly when the dividing cells contact or "overgrow" one another. Seen in this light, experimenters' manipulations *block* stem cells innate tendency to differentiate, rather than push them toward differentiation.

of the cultured cell lines were plated into a range of different cell culture environments, each designed to encourage development along a particular pathway. That is, samples of each cell line were encouraged to differentiate into a collection of specific cell types, using cell culture environments calibrated to each cell type (neural, muscle, blood, bone, skin, etc.). Thomson et al. did not test their cultured cells against every human cell type (that would have required more than 200 different culture environments and comparisons) but a representative range of major ones, for which cell culture conditions had been discovered. Because it is onerous to perform these experiments for every major developmental pathway in the human body (not to mention we are ignorant of many of them), it has become standard to use proxies for the full range of cell types to test cell lines for pluripotency (see the later discussion). After 1998, Thomson et al.'s method, like the original spleen colony assay, was adapted and modified to produce many variants on the original theme: many subtly different ways of growing stem cells derived from embryos, each yield-ing a slightly different ESC line.[70]

Making Pluripotent Stem Cells

The last exemplary experiment I will discuss is "reprogramming," which pro-duces induced pluripotent stem cells (iPSC; Takahashi and Yamanaka 2006). The name is a good description: stem cells of this variety are "induced" because they are produced by experimental manipulation, "pluripotent" because, like embry-onic stem cells, they can differentiate into all the main cell types of the organismal body. iPSC were invented to exhibit pluripotency without being derived from an early embryo, as a way to circumvent ethical and political controversies over the use of human embryos for research.[71] Shinya Yamanaka's driving idea was to transform differentiated mammalian cells – mouse fetal skin cells, in his experi-ments, but many cell types will do – into cells resembling ESC. In other respects, his method was very like that for hESC and other cultured stem cell lines. The innovative step was transforming differentiated cells to cells resembling ESC. If ESC culture "freezes" cell development at an embryonic stage, reprogramming reverses it; unwinding developmental pathways to "push" cells back to an

---

[70] A few months after Thomson et al.'s landmark result, a second group, led by John Gearhardt at Johns Hopkins, reported another human embryo-derived stem cell: the germline stem cell, or GSC, derived from primordial germ cells of human embryos (Shamblott et al. 1998). Other variants, like epiblast stem cells, followed later. More recently, "expanded potential stem cells" (EPSC) have been grown from pig and human embryos – these stem cells correspond to the earliest, totipotent embryonic cells.

[71] See Stem Cell Reports (2020). This did not happen – in part because of the diversifying, ramifying pattern of experiments and models in stem cell research (discussed later).

embryo-like state. To accomplish this, Yamanaka's team used a molecular genetic approach, which amounted to a form of bioengineering.[72]

Reprogramming begins with differentiated cells in culture: mouse skin cells, in Yamanaka's original experiments. To the culture medium, Yamanaka added a retrovirus with four mammalian genes spliced into its genetic material.[73] These genes encoded *transcription factors* (TF): proteins that specifically bind to particular sequences of DNA, affecting the expression of genes nearby on the chromosome.[74] Altering the expression of one TF gene can affect the expression of dozens, even hundreds, of other genes – including other TF genes. So TF proteins act as "switches" coordinating large-scale patterns of gene expression within a cell. Retroviral infection imprecisely integrates the four TF genes (along with other viral genes) into the genome of cultured cells; a transplant at the genetic level, but without the precise stitching of organ transplants or gene editing. Mammalian gene expression is intricately regulated, so random insertion of genes somewhere in the genome rarely results in the expression of the new additions. But, as is typical for molecular biology, the experiment needs to work only once among thousands or even millions of cells. Yamanaka's group also engineered a way to select cells expressing virally inserted genes. After a few weeks, a few of the cultured mouse skin cells (about 0.5 percent in early experiments) transformed to resemble ESC, morphologically and in their reproductive and developmental capacities. From this point, Yamanaka's method coincided with ESC experiments: "reprogrammed" cells were matched to ESC in their molecular, biochemical, and cellular traits, as well as self-renewal and differentiation abilities. The result was a collection of iPSC lines resembling mouse ESC, derived from ordinary cultured skin cells (Takahashi and Yamanaka 2006). The first human iPSC lines were made a year later by the same group (Takahashi et al 2007). An avalanche of iPSC followed, made in laboratories worldwide from many different kinds of starting cells using various culture conditions.[75]

As noted, the key innovation in this experimental strategy was adding the four TF. The interconnected mechanisms of gene expression that determine cell phenotype are not well understood. So, again, trial and error was needed, to find

---

[72] Molecular and genetic traits are used to characterize HSC and ESC, so those tools are not entirely absent in the other exemplars. But molecular genetic methods are distinctively central to the iPSC experimental strategy.

[73] The four TF – Oct3/4, Sox2, Klf4, and c-Myc – now referred to as the "Yamanaka factors."

[74] "Gene expression" is a catchall term for a sequence of DNA being transcribed to RNA and subsequently having some effect within a cell and/or on its environment (including other cells).

[75] Nishizawa et al. (2016) provide a recent overview, matching existing iPSC lines with specific applications based on subtle differences in differentiation potential and molecular characterization.

the right combination of TF to "capture pluripotency."[76] Yamanaka's team began by making a list of genes/proteins plausibly implicated in maintaining cells in an embryo-like state, building primarily but not exclusively on ESC experiments.[77] This literature search yielded twenty-four candidate factors. Yamanaka's group combined them all onto a retroviral vector and successfully induced pluripotency in (a very few) mouse skin cells. Next, they added each factor to cells individually – with no results. This showed that the factors needed to interact; no single one of them could induce pluripotency. Yamanaka's team then *subtracted* each candidate individually, trying all possible different combinations of twenty-three factors. Ten of these yielded no results, indicating that the missing factor in each case is necessary. Those ten were combined and together induced pluripotency *more efficiently* than the original set of twenty-four candidates. (This indicates nonlinear genetic interactions among TF.) The team performed individual subtraction experiments for each of the ten, identifying four as necessary to induce pluripotency.[78] Those four together, and no others, induced pluripotency about effectively as the ten: these were the four published "Yamanaka factors." Their stepwise experimental design cleverly selected a small interacting set of factors, without advanced knowledge of those interactions or how the factors worked.

Once the inducing factors were identified, Yamanaka's experimental method was rapidly transferred to many laboratories across the world. Modifications of the original method led to a proliferation of variants, and consequently to an explosive radiation of iPSC lines. Rather than a single standard procedure, reprogramming experiments are variants on a general theme. These variants collectively produce myriad iPSC lines.[79] In recognition of this achievement and consequent insights into biological development, Yamanaka received the 2012 Nobel Prize in Physiology or Medicine.[80] Characteristically for stem cell

---

[76] A phrase taken from Ying and Smith (2017). In this exemplar, trial and error played out on a smaller scale – within one laboratory over months, not worldwide over years. The basic method was adapted from groundbreaking experiments transforming skin to muscle cells in culture.

[77] Basically, they looked for genes that were highly expressed (meaning lots of RNA or protein encoded by those genes was found) or necessary to maintain a pluripotent state (meaning if those genes were deleted or not expressed, cells differentiated). A key hypothesis was that "the factors that play important roles in the maintenance of ES cell identity also play pivotal roles in the induction of pluripotency of somatic cells" (Takahashi and Yamanaka 2006, 663).

[78] This is also nonlinear: the ten essential factors of twenty-four are not the same as the four essential factors of ten, although the two sets are about equally effective (about 0.5 percent of cells).

[79] Reviewed in Maherali and Hochedlinger (2008), Wilmut et al. (2011), and Brumbaugh et al. (2019). Because of their genomic basis, iPSC are a tool for precision medicine; personalized pluripotent cell lines can be made from just about any cell of a person's body (see Section 4).

[80] This prize was shared with John Gurdon, who fifty years earlier successfully reprogrammed frog nuclei using egg cytoplasm.

biology, there is more experimental uptake of reprogramming than (published) reflection on its broader implications. Concerning the latter, Christina Brandt (2012) considers the term "reprogramming," which prima facie suggests a pre-existing plan or program for development (which experimenters *re*-program by adding TF).[81] Despite this impression, for mammalian cells "programming" refers only to the idea that a cell's developmental state and potential (i.e., its position within a developmental lineage) depend on an underlying arrangement of molecular "circuitry" (DNA, RNA, protein, and small molecules). "Reprogramming" refers to the process of changing a cell's location in a developmental lineage, by manipulating the underlying molecular network. Indeed, the experimental success of reprogramming *undermines* the idea of a fixed program for mammalian cell development.

Reflections on Method

Several general points emerge from this survey of exemplary experiments in stem cell research. First, the experiments are epistemically sophisticated, involving considerable strategy and nuance. Philosophers, drawn to abstraction, tend to underappreciate sciences that work in concrete media. I hope to have indicated some of the ingenuity and effectiveness of stem cell experiments as means of producing scientific knowledge. Much of this knowledge is embodied in concrete cell lines and methods of finding stem cells within organisms. Trial and error is involved in these experiments, but delicately deployed as part of a nuanced experimental strategy. Key epistemic moments in each exemplar involve frame-shifting between individual research groups and a wider experimenting community.[82] Indeed, the practices involved in designing experimental methods are as philosophically rich as those for constructing abstract theoretical models – and can be fruitfully conceived as practices of concrete model construction.[83] The epistemology of experimental fields like stem cell research richly repays sustained philosophical attention.

Second, and more specifically, stem cell experiments are *fruitful* in a way that mirrors their targets' proliferative ability. Although each exemplar began as one specific experimental method innovated by a single research group, they soon diffracted into many variants, taken up and modified by research groups across the world. The proliferation of methods for finding, growing, and making stem cells yields a profusion of stem cell varieties. The labels "HSC," "ESC," and "iPSC" do not name one thing but many. The latter two are ever-ramifying collections of

---

[81] Brandt explicates reprogramming as changing a cell's "internal time" (2012, 55). See also Fagan (2015b).

[82] See Fagan (2007, 2010, 2011) for more detail.

[83] I return to this point at the end of the section.

cultured stem cell lines.[84] The first is a set partly overlapping subpopulations of cells isolated by a cluster of methods continuously refreshed by technological innovation. Furthermore, there are many other varieties of stem cells than those three. These others can be classified according to the method used to identify them (each of which, when realized, specifies the variables of the abstract model). Stem cells found in organisms' bodies are for the most part identified and characterized via methods descended from the pioneering HSC experiments. The basic approach for finding blood stem cells in bone marrow is applied to other tissues: muscle, brain, skin, eye, and many more.[85] Variations on the ESC method that begin with embryos of slightly different stages and/or location of cell extraction produce cultured stem cell lines with subtly different abilities than ESC. So methods for identifying HSC and ESC are templates for identifying other tissue-specific and pluripotent stem cells.[86] Although stem cells are sometimes classified as "adult or embryonic," depending on the source organism's developmental stage, the distinction between in vivo (found) and in vitro (grown or made) is more methodologically fundamental. All the main varieties of stem cells are identified and characterized by experiments that can be traced to one of these exemplars.

A third insight follows. With the efflorescence of variants, each experimental strategy can be represented as a general schema that can be filled out in many different ways. To find stem cells, the basic design is as follows: (1) extract cells from some tissue or organ (e.g., bone marrow), (2) sort cells into subpopulations according to distinguishing traits, (3) move cells to a new environment (e.g., an irradiated mouse, cell culture for B or T cells, fluid suspension), and (4) after some time-interval, test each subpopulation's self-renewal and differentiation ability. To grow stem cells in artificial culture, the procedure is as follows: (1) extract cells from some location within an early mammalian embryo; (2) move cells to an artificial environment encouraging self-renewal, selecting for cells with that capacity and other distinguishing traits; (3) move cells to a set of new environments encouraging differentiation along relevant pathways; and (4)

---

[84] Although iPSC were invented to bypass ethical challenges to hESC, the former has not replaced the latter. Instead, research including both has flourished and generated more lines of each (see Kobold et al. 2015 for an inclusive overview of trends based on bibliometrics).

[85] And for other stem cells in bone marrow, notably bone-producing stem cells. Hair follicle stem cells are very well-characterized and part of an accessible and continuous regeneration system.

[86] Patterns of variation in methods for identifying stem cells are also revealing. There is enormous profusion of ESC and iPSC cell lines, but not many different varieties of cultured pluripotent stem cell (e.g., germline, epiblast). The pattern is reversed for tissue-specific stem cells: there are multiple variants of HSC, but not the wild proliferation as for ESC and iPSC. Instead, there are many different varieties of tissue-specific stem cell, corresponding to different parts of an organism's body (tissues, organs, cell types). A first thought as to why this is the case is that each experimental approach to identifying stem cells is aimed at "grasping" all of organismal development. Pluripotent stem cells enable this grasp via their developmental potential, while tissue-specific stem cells do so only collectively.

after some time-interval, test for differentiation ability by comparing developmental products with in vivo counterparts. For reprogramming, different combinations of starting cell population (skin, gametes, blood, muscle, etc.), TF (e.g., the four Yamanaka factors, the "core pluripotency network" of Oct4, Sox2, and Nanog, and many more), delivery method (e.g., retrovirus, gene editing), and culture conditions (presence or absence of specific biochemical factors) specify variants of a general method for making stem cells. Different combinations of values of these variables are more or less effective at producing iPSC with specific developmental capacities, such as the ability to produce neurons, blood, or cardiac muscle in vitro. Otherwise, this method has the same basic steps as for growing stem cells, except for the first: (1) extract cells from some location within a developed organism.

Putting these schemata together, a general pattern emerges – a common basic design for experiments aiming to identify and characterize stem cells. First, cells are extracted from some organismal source. Cells are then moved to a new environment encouraging self-renewal, and characters thought to distinguish stem cells (size, shape, gene expression pattern, etc.) are measured. The duration of this step specifies the extent of self-renewal. Differentiation potential is specified by realizing measured cells' developmental capacities in a permissive environment or range of environments and comparing results with products of normal development in the relevant species. Variables that fill out this general schema dovetail with the abstract model presented in Section 2 (Figure 6): organismal source S, interval of self-renewal n, stem cell characteristics C, a developmental hierarchy terminating in some range of specialized, differentiated cells, which contributes to some organismal product P. Each variety (or subpopulation, or line) of stem cells corresponds to a different combination of values of variables in the abstract model. Those values are specified by the details of the experimental method used to identify that particular stem cell variety, subpopulation, or line. Because experiments with this basic design are the only way stem cells are identified in practice, the abstract model schematically covers the immense variety of stem cells, while highlighting their experimental ground and origins. The lineage conception of stem cells extends to experimental methods for identifying bits of biological reality that fit that concept. This completes the mediating connection between abstract philosophical analysis and concrete stem cell experiments, begun in the previous section.

To illustrate, each of the exemplars discussed here amounts to a class of specifications of the abstract stem cell model. HSC are derived from the bone marrow of an adult mouse (Condition 1) and inferred to self-renew with respect to the set of characters used to distinguish them from other bone marrow cells (Condition 2), for an interval sufficient to regenerate an irradiated mouse's blood and immune system (Condition 4). HSC so characterized are the origin of the blood and immune cell developmental hierarchy, the structure of which is partly specified by the mature blood and immune cells measured in the "read-out" (Condition 3). Similarly, hESC are derived from the inner cell mass of a human blastocyst (Condition 1) and are capable of initiating a reproductive cell lineage via unlimited cell division, preserving sameness with respect to a suite of cell characters (Condition 2).[87] ESC so characterized can initiate a developmental hierarchy with termini corresponding to all the major cell types in the human body (Conditions 3 and 4). iPSC are derived from any type of differentiated cell from a post-embryonic mouse or human (Condition 1) and initiate an unlimited reproductive cell lineage with respect to the same characters as ESC (Condition 2). Also, like ESC, iPSC can originate many different developmental hierarchies of varying structure, with end points corresponding to all cell types of a mature organism (Conditions 3 and 4).

As the exemplars show, the abstract model is not fully specified by each stem cell experiment. In particular, the exemplary experiments do not uniquely specify a lineage tree structure or cell differentiation hierarchy (see Section 2). Some experiments do specify the number of hierarchical stages, via the timing of measuring cell characters or other ways of distinguishing different lineage positions in the lineage. Tests of developmental potential specify the number of lineage termini. And the lineage origin is always specified – the stem cell that is the focus of the experiment. Arrangement of branch-points, though, is often underdetermined. Sometimes background knowledge about the cellular system in question fills in the gaps. But in many cases, the stem cell identified by a particular experimental method corresponds to a set of a possible lineage tree structure, constrained but not uniquely specified by the methods used and assumptions made. In these cases, it is helpful to think of stem cells as defined by the terms of the abstract model, corresponding to all possible lineage tree structures compatible with experiments and their results.

The stem cell identified by a particular experiment corresponds to some region of *lineage tree space* – some set of possible tree topologies. The notion of *tree space* is a good way to visualize the field of stem cell experiments and resulting varieties of stem cell, which accommodate all the cross-cutting and overlap among those diverse varieties in scientific practice.

---

[87] In the original experiments, these are rapid division in culture, lack of specialized traits, flat round shape, large nuclei surrounded by correlatively thin cytoplasm, prominent nucleoli, high telomerase activity, and high expression of particular genes.

An "Uncertainty Principle" for Stem Cells

Diversity is a prevailing theme in stem cell experiments and their results. Why might this be? Why does the field not converge on a few clearly distinguished stem cell types, rather than finding, growing, or making myriad varieties that overlap and cross-cut in their traits and abilities? One reason is that experimentally identifying a cellular entity that matches the stem cell concept is inherently *uncertain*. In philosophy of science, uncertainty is typically conceptualized in terms of evidence for propositions (hypotheses or theories) and as a factor in decision-making (primarily, whether or not to accept a hypothesis). That traditional approach does not sit entirely comfortably with my focus on models and experiments. So the way I explicate uncertainty about stem cells is somewhat nontraditional. The thesis of this section is that broad swathes of stem cell research practice can be understood as efforts to manage, circumvent, or minimize the inherent uncertainty of identifying a stem cell. Like the abstract model of Section 2, the "stem cell uncertainty principle" is a philosopher's construct, grounded on stem cell research practices and aimed at conveying their key features to a nonexpert audience.

It is important to be clear what "stem cell uncertainty" is about. It is not about the idea or concept of a stem cell, which has been part of life science since the mid-nineteenth century. Nor is the existence of stem cell lines or subpopulations of cells extracted from organisms in doubt – those concrete models are living in laboratories all over the world, with their targets invisibly maintaining our bodies. What is uncertain is identifying *a single cell* – a biological individual, bounded from its environment by an enclosing membrane – as *a stem cell* with specific capacities for self-renewal and differentiation. The duality of the idea of a stem cell poses a challenge for that experimental aim. To see why this is so, consider, again, self-renewal and differentiation. Both are processes in which *an individual stem cell* passes out of existence. To self-renew is to produce either one or two cells resembling the parent, which ceases to exist upon splitting. To differentiate is to become something other than a stem cell; a stem cell either transforms into something else or divides to produce offspring exhibiting that transformation. Stem cells, by definition, give way to their descendants.

Now consider an experiment designed to test whether some individual cell is a stem cell. To show that it is (or is not), that cell's capacities for self-renewal and differentiation must be measured. To be measured, those capacities must be realized. But upon such realization, that individual cell is no more.[88] This creates

---

[88] Experiments to identify and characterize stem cells are thus "destructive" of their object. A similar situation arises in nanotechnology (see Fagan 2019).

three distinct evidential problems.[89] First, it is not possible to measure both self-renewal and differentiation potential for a single cell. To test a cell's differentiation potential, that cell has to be in an environment *conducive to* differentiation. To test a cell's self-renewal ability, the cell is placed in an environment that *blocks* differentiation. For both, that cell's descendants need to be measured. But there's no way to generate descendants from one cell in two environments, when the cell is gone after generating descendants. So both stem cell capacities cannot be measured for a single cell. Because stem cells are defined as having both capacities, it follows that no individual cell be identified as a stem cell. Nor does it help to separate the two capacities. Self-renewal, recall, occurs just in case parent and offspring cells are the same with respect to some characters of interest, for some time-interval of interest. A self-renewing stem cell divides to produce one or two offspring that are also stem cells, with the same ability to self-renew as the parent (and the same differentiation potential). But the offspring stem cell's ability to self-renew is revealed, in turn, by *its* offspring having the same capacities – and so on. Data showing that a stem cell self-renews are always one generation in the future. So experimental proof of self-renewal for an individual stem cell is infinitely deferred. Differentiation potential is revealed by placing a cell in an environment conducive to differentiation and then measuring its descendants to see if they exhibit the features of some range of differentiated, specialized cell types. However, there is no generic differentiation environment. Instead, different kinds of specialized cells are elicited by different environments (cell niches). A single cell can be placed in at most one such environment. Experiments cannot tell us what a cell's descendants would be like in a different range of environments.

For all three reasons, claims that any single cell is a stem cell are inevitably uncertain. It is impossible, strictly speaking, to experimentally demonstrate that any individual cell is a stem cell. This amounts to a kind of "uncertainty principle" for stem cell biology.[90]

How then can stem cell experiments succeed? Is the whole enterprise a sham? No. There are shams involving stem cells, discussed in Section 4. But the field's central experiments are not among them. This "uncertainty principle" does not cast a blanket of doubt over all stem cell science. It is a conceptual constraint that clarifies what stem cell experiments can and

---

[89] More detailed versions of this argument appear in Fagan (2013a, 2013b, 2015a).

[90] Some stem cell researchers have noticed the problem and explicitly connect it to Heisenberg's uncertainty principle in physics (see Potten and Loeffler 1990, p. 1009, for a very clear statement). Although there are some intriguing parallels, analogies between theoretical physics and experiment-driven stem cell biology are limited and should not be pushed too far. I discuss this issue in detail in Fagan (2016a).

cannot show. They cannot identify and characterize an individual stem cell as such. To do that, experimenters would need a population of identical cells (a clone). That would allow tests of what amounts to the same cell across different environments, conducive to self-renewal and to a range of differentiation pathways, resolving the evidential problems described previously.[91] For example, if we knew the characteristics of all and only mouse HSC (e.g., a combination of cell surface molecules such as $CD32^-/CD127^-/CD117^{hi}/Sca1^+/CD135^{hi}$), then we could isolate exactly those cells. We could then transplant identical single cells into different environments and measure their progeny to detect self-renewal (i.e., offspring with the exact same combination of cell surface molecules) and differentiation (i.e., descendants that match one or more specialized cell types). However, if we knew the mouse HSC profile, then we would not be trying to identify and characterize that variety of stem cells – because we would already have that knowledge. But how to acquire it in the first place? Experimenters seem trapped in a vicious circle: to interpret the results of a stem cell experiment as identifying any stem cell variety presupposes that experiment's success.[92]

The solution is to give up the idea of a "crucial stem cell experiment" – a single experimental method that decisively identifies and characterizes a variety of stem cell, full stop. Any particular experiment to identify a variety of stem cells rests on the assumption that the candidate stem cell population is homogeneous. Homogeneity of the population is routinely measured in a stem cell experiment – for a particular set of traits. We cannot know in advance, though, that those traits are the right ones that pick out stem cells of that variety. For any stem cell experiment, the homogeneity assumption is a necessary working hypothesis – always provisional, always uncertain. (It is obviously risky to presume perfect homogeneity for a population of cells whose essential feature is transformation.) As new cell traits are discovered and made accessible to measurement, the assumption of homogeneity must be continually reassessed and revised. Characterizations of stem cell varieties are therefore provisional, and become obsolete when new characters and environments are introduced. That is the lesson of stem cell uncertainty: what an experiment identifies as a stem cell is *relative* to the assumption that its test population is homogeneous, that they are "the same cell" with respect to the set of traits and environments used in that experiment. Stem cell experiments identify and characterize stem cells only *relative to particular experimental methods*.

---

[91] Infinitely deferred self-renewal is avoided by defining self-renewal with respect to the characters shared by the set of identical replicate cells (clones), not by stem cell capacities.

[92] This is related to the famous "experimenter's regress" problem.

Because of this *experimental relativity* stem cell research proceeds by multiplying experiments and trying out many variations on a basic method. That is how knowledge of stem cells accumulates – not by refining general patterns or formulas, but by multiplying experimental contexts in which stem cells are rigorously identified and characterized. The "uncertainty principle" does not mean that we cannot know about stem cells. It does mean that this knowledge is context-dependent and provisional – it depends on assumptions about the experiments that need to be updated and re-assessed as methods change in response to technological advances. Experimental relativity is also a facet of stem cells' own context-dependence (Section 2). It has long been recognized that the "stem cell phenotype appears to be by no means hard-wired" but rather "the behavior of stem cells is controlled by the microenvironment" (Potten and Lajtha 1982, 454; see also Mesa et al 2015, Yin et al 2016). More than fifty years of stem cell experiments show that stem cell identity is exquisitely sensitive to features of the local microenvironment, in vivo and in vitro. Indeed, manipulating this context-dependence is the basic design principle of all stem cell experiments: candidate stem cells are moved to a new environment, and their or their descendants' characteristics in that context are measured. Coaxing stem cells to differentiate in various specific ways by manipulating their environment's geometry, biochemical composition, or cellular makeup is at the heart of the field. Multiplying those experimental setups is how we progressively acquire knowledge of stem cells and their abilities.

Alongside the proliferation of experimental contexts, "stem cell uncertainty" clarifies other important features of stem cell research today. For decades, stem cell biologists have looked to experimental techniques to manage uncertainties that follow from the stem cell concept. The evidential constraints underlying the uncertainty principle pose an enduring challenge that researchers aim to meet and a bar for judging scientific progress. One robust strategy is "the single-cell standard" (Fagan 2013a, b): the experimental ideal of a single cell in a controlled environment, with all relevant signals taken into account, so results reflect all and only the reproductive and developmental output of that single starting cell. Measured stem cell capacities can then be unambiguously attributed to that cell *in that environment*. This ideal single-cell experiment is an important standard for progress and success in stem cell biology – an organizing methodological principle that continues to shape the field. This dates back to the earliest stem cell experiments: Till and McCulloch (1961) announced their spleen colony assay as "a single-cell technique" (213). Much of the stem cell research from 1961 to the present can be seen as efforts to meet this same standard. For example, the "gold standard" for demonstrating stem cell abilities is a single-cell transplant leading to long-term reconstitution of a tissue or organ (Melton 2013). Technical innovations that increase experimenters' ability to measure and track single cells more closely approximating the ideal single-cell standard. Post-genomic and imaging technologies that enhance our ability to isolate or track single cells have been quickly adopted by stem cell biologists and reported as advances in the field. Virally inserted "barcodes" offer prospects for tracing cell lineages, enabling researchers to track the progeny of a single cell and their developmental trajectories. The most recent iteration is single-cell lineage tracing, *Science* magazine's 2018 Breakthrough of the Year (Pennisi 2018). Advances in RNA sequencing, in vivo visualization, and computational biology allow biologists to track single cells in organismal development and reconstruct their lineage relationships. Although examination of this confluence of methods is beyond the scope of this short essay, its significance for stem cell biology is fruitfully understood as the latest important application of the field's long-held single-cell standard. Instead of progress through refinements of theories and models, stem cell biology progresses through technological innovations that improve our access to the reproductive and developmental activities of single cells.

The uncertainty principle articulates a constraint within which these experiments operate, thereby making intelligible a number of general features of stem cell research. In addition to the single-cell standard, these include long-running debates over the developmental potential of stem cells from different sources, scarcity of generalizations, and the notoriously rapid turnover in stem cell hypotheses and terminology. Substantive claims about stem cells are provisional and rapidly become obsolete. Generalizations, laws, or rules about stem cell capacities are very hard to come by. Indeed, there are (as yet) no general rules for cell development and other stem cell phenomena. Knowledge in the field takes another form, distributed among diverse experimental methods and models. The inherent uncertainty of stem cell experiments and consequent experimental relativity of claims about stem cells allows for prolonged debate between laboratories about the capacities of stem cells from particular organismal sources (e.g., HSC, "mesenchymal stem cells," etc.). The provisional working hypothesis that some variety of stem cells is already in hand is easily confused with robust scientific fact, leading to an inflated view of our epistemic situation vis-à-vis stem cells. Stem cell uncertainty and its consequences offer a corrective to such premature optimism and the associated tendency to over-hype the prospects for stem cell research. In this way, my philosophical account presents a more realistic and informative view of the field, for audiences outside the community of stem cell researchers.

A Fabric of Generative Models

I began this section by observing that stem cell biology lacks the abstract representations (laws and theories) that philosophers are trained to look for. Instead, the field is comprised of a welter of experiments. But all is not "blooming, buzzing confusion."[93] The single-cell standard is widely shared, responding to a challenge that transcends differences across methods. The diversity of stem cell experiments and their products is a response to experimental relativity. The stem cell uncertainty principle thus marks out an epistemic core of the field, its far-flung diversity of experiments notwithstanding. To conclude this section, I return to the issue of knowledge. How is knowledge of stem cells and related phenomena produced out of myriad different experiments? One pattern emerges clearly from the discussion so far: stem cell experiments radiate from a point of innovation. That is, a key experimental system, like Till and McCulloch's assay, is applied with modifications so as to yield a burst of related-yet-different stem cells. The task is then to discover the abilities, uses, and fine distinctions between these various experimental products. The pattern is exhibited across HSC variants, among the various tissue-specific stem cells, cancer stem cells, and diverse pluripotent stem cell lines. Even though Thomson's original hESC grow and divide continuously, the stem cell field has produced many, many more lines (accompanied by ethical controversy and associated funding challenges). And researchers made not only more hESC lines but also many other kinds of cultured stem cells.[94]

What are the implications of this pattern, for scientific knowledge of stem cells? A phenomenon of interest – mammalian development, especially *human* development – is represented by a sprawling family of concrete models: cultured stem cell lines and (relatively) pure populations of stem cells found in organisms. These models are related to one another and to stem cells found in vivo by a thicket of analogies: comparisons and contrasts in molecular and cellular traits, as well as developmental abilities. These relationships are revealed by the details of experimental methods used to construct each model. So experiments tell us how different stem cell lines and populations are related, and then how each represents some small part of mammalian development. In summary, stem cell experiments and their results comprise an interconnected fabric of models, from which insights about development (especially human development) emerge piecemeal.[95] The idea of a stem cell is thus, in scientific

---

[93] This phrase is from James (1890, 462).

[94] "An observable tendency is the ad hoc establishment of ever more hPSC lines tailored for a specific application in many individual laboratories" (Kurtz et al. 2018, 2).

[95] Fagan (2013a, 2016b) argue this claim in more detail.

practice, intricately interwoven with concrete stem cell experiments, and with our emerging knowledge of organismal development. Human development (indeed, all forms of mammalian development) is an extremely complex phenomenon, which no single, tractable model can adequately represent. Stem cells are a powerful tool for modeling aspects of development and also for generating other models. In this way, stem cells' scientific modeling role mimics their developmental capacities.

One aspect of this dual role is *developmental versatility*. The term refers to the multiple ways that stem cells contribute to organismal development, not just producing specialized cell types but other forms of organismal organization (Fagan 2017, 2018). Developmental versatility is revealed, unsurprisingly, by experiments that characterize differentiation potential. As noted previously, those experiments often use proxies instead of directly testing whether a given stem cell (population) can produce all the major cell types of an organism's body. Two classic proxies are embryoid bodies and teratomas. These are three-dimensional structures derived from cultured pluripotent stem cells, which mimic early (human) embryos. Pluripotent stem cells typically grow on a two-dimensional solid surface (a culture dish or layer of "feeder" cells), on which they divide to form colonies (self-renewal). Embryoid bodies are produced by suspending cultured pluripotent stem cells in fluid. Floating free from their two-dimensional colonies, stem cells organize into tiny spheres, about 0.1–0.2 millimeters in diameter. Each sphere consists of an inner core of undifferentiated cells wrapped in a layer of more differentiated cells. These simple cellular structures resemble an early stage of mammalian embryonic development; specifically, the formation of germ layers, with inner and outer cells committed to different fates.[96] The ability to form embryoid bodies with different fates for inner and outer cell layers is a quick test of pluripotency. Teratomas are produced by injecting human stem cells into a strain of inbred mouse that does not reject the cells but allows them to grow as tumors within the mice. These tumors consist of diverse specialized cells and tissues, jumbled together to make an upsetting mass of misplaced structures such as teeth, nerves, and hair. Tumors are collected, dissected, and histologically analyzed to determine the range of cell types that develop from injected stem cells. This method is actually a relic of early research on stem cell lines, which were transmissible forms of cancer (Andrews 2002). If tumors contain cells from

---

[96] The three classic germ layers – ectoderm, endoderm, and mesoderm – result from early differentiation branch points, corresponding to separate developmental pathways leading to clusters of tissues and organs. Separation of germ layers is thus a key differentiation event.

all three germ layers, however disorganized, then scientists conclude the transplanted stem cells are pluripotent.

We can now reconsider the most straightforward test stem cell pluripotency: "directed differentiation" experiments. In these experiments, samples of a stem cell line are placed in two-dimensional cultures with biochemical factors encouraging differentiation along various distinct pathways. For pluripotent stem cells, all the main pathways should be represented: blood, bone, liver, neurons, and so on. For tissue-specific stem cells, a narrower range of outcomes is relevant. For any directed differentiation experiment, the result is a set of sheet-like populations of particular specialized cell types, demonstrating that a given stem cell's potential includes those cell types. The differentiation process is fragmented into distinct pathways – there's no higher-level entity with multiple cell types, like an embryoid body or teratoma. Considered collectively, directed differentiation experiments are a kind of limit case; thoroughly disarticulated cell development lacking all aspects of organismal organization. Now consider the various methods of testing stem cell developmental potential and the new proxies for normal human development. Normal development, the unmanipulated case, is what we want to understand. Embryoid body formation is a simple, minimal model of that process. Teratoma formation is its pathological counterpart, disorganized in multiple respects. Direct differentiation is a purely cell-level process, as though the organism were fragmented into homogeneous sheets of distinct cell types that, properly organized, make up its functioning body. Each offers a distinct window on the transformation from cell to organism, revealing as an aspect of organismal organization (Table 2).

Two new kinds of stem cell experiments reveal further aspects, in the form of organoids and embryo-like structures. These are models of development in mammalian organs and early embryos, respectively. Organoids are a new kind of biological entity, produced by in vitro developmental processes that (1) are initiated by stem cells, (2) mimic or "recapitulate" the cell-level processes of normal development, and (3) produce multiple organ-specific cell types that are organized to perform at least one organ-specific function (e.g., neural activity, excretion, contraction). These artificial, simplified proto-organs are made by seeding stem cells onto a synthetic three-dimensional scaffold and letting them grow in biochemical conditions thought to mimic those of normal organ development (Lancaster and Knoblich 2014, Science 2019 Special Section). In their new biochemical environment and with changed spatial relations to one another, pluripotent stem cells differentiate to yield more mature cells that then interact to produce an approximation of an organ: optic cup, brain, intestine, liver, kidney, stomach, pancreas, and more. Importantly, the approximation has to be both

Table 2 Stem cell experiments and developmental versatility.

Experiment	Mode of development	Aspect of organization
n/a	normal	all: complex, robust
embryoid body	simplified	generic cell types, simple 3d structure
teratoma	pathological	specialized cell types, complex 3D structure, non-robust
organoid	organ-like	specialized cell types, complex 3D structure, non-robust
embryo-like strucrure	embryo-like	generic cell types, simple 2d structure, polarity
directed differentiation	cellular	specialized cell types

structural and functional: neural activity in "mini-brains," digestive enzymes in gut organoids, and so on. The latest in vitro stem cell derivatives are embryo-like structures – which are exactly what the name suggests. There are several varieties already; more are probably on the way.[97] The simplest method of producing embryo-like structures is to impose simple spatial constraints on growing ESC cultures and adding a molecule involved in normal differentiation.[98] For human embryonic stem cells, that molecule is BMP4, and the spatial constraints are small circles. In this environment, hESC "replicat[e] embryonic spatial ordering in vitro" (Warmflash et al 2014, 847). They differentiate in a "center-to-periphery" pattern, with "stemness" remaining highest at the center, declining toward the periphery. Of course, this simple radial structure does not resemble an embryo anatomically. But embryo-like structures do exhibit several aspects of embryonic organization, including robust timing and order of appearance of cell types, a reproducible spatial arrangement of germ layers, and polarity along the radial axis. Interestingly, the stage of normal development this resembles is not

---

[97] Earlier embryonic development is simulated by "blastoids," made by combining ESC with trophectoderm stem cells – the two fates after the original differentiation event. "Neuraloids" represent later stages, modeling the process of early neural development – both cells and morphological structures.

[98] Directed differentiation of cultured stem cells is controlled (as far as we understand the process) by signaling molecules in a particular temporal order and concentration; with a very intricate orchestration of developmental stages forming a branching lineage tree (differentiation hierarchy).

the very early embryo (as for embryoid bodies) but a later one: gastrulation. In normal, unmanipulated development, gastrulation is the stage when cells are committed to form a single multicellular organism, the three germ layers are distinguished, and generic cell fates are fixed.[99] (The procedures for mouse ESC are somewhat different, with evidence of germ-layers in the embryo-like structures, but the basic experimental approach is the same.)

Each of these stem cell derivatives – embryoid bodies, teratomas, cultured cell types, organoids, embryo-like structures – highlights different aspects of the organization involved in generating a multicellular organism from undifferentiated cells. This goes beyond the standard notion of stem cell potential, defined in terms of the range of organismal cell types a given stem cell can produce (Section 2). Collectively, pluripotency experiments show that stem cells can realize different processes or modes of organismal development. "Developmental versatility" refers to this extension of stem cells' differentiation potential. Experimental methods that realize this reveal continuity between undifferentiated cell colonies and mature multicellular organisms, through intermediates like embryoid bodies and teratomas. The various experimental products (directed differentiation cell populations, embryoid bodies, teratomas, organoids, embryo-like structures) all realize stem cells' developmental capacities, but in different ways. These correspond to various modes of differentiation involved in organismal development. At least six such modes can be distinguished, each highlighting different aspects of the process of generating a multicellular organism from undifferentiated cells: simple, dispersed, cellular, organ-like, and embryo-like; all integrated in the in vivo process of normal organismal development. No single stem cell derivative captures every significant aspect of human development. Instead, the different models complement one another, highlighting different aspects of the in vivo process. In this way, stem cells' context-dependence is deployed in an experimental strategy to reveal piecemeal the diverse factors involved in generating organismal organization from a stem cell.

In sum, knowledge arises in stem cell research not only from individual experimental systems but also from a fabric of interrelated models. The models are of biological development – that complex phenomenon is their common target. The purpose is knowledge or understanding. The agent is the entire community of stem cell research – or, at least, much of that community

---

[99] Around this time, in human embryos, the primitive streak forms, and there's a long-standing ethical restriction against letting human embryos develop past this point (the fourteen-day rule). Embryo-like structures are prompting re-examination of that rule. The ISSCR is currently formulating guidelines for work on embryo-like structures, to be announced in 2021 (details at www.isscr.org). See Hyun et al. (2020) for a summary of mouse and human embryo models to date.

and all the research groups working on stem cell models of biological development in humans and other species.[100] Stem cell models, because of their generative nature, are well-suited to this kind of epistemic organization: a fabric of interrelated models for understanding a complex phenomenon. There is much more to analyze about this epistemic organization and the forms of knowledge that emerge from it. Stem cell research is radically decentralized, with no standard definition, nomenclature, or classification principles of its eponymous focus. This social epistemic organization, plausibly, has consequences for the knowledge produced. So a deeper look at the social organization of stem cell experiments would be valuable, both scientifically and philosophically. Most stem cell experiments take place within one or a few laboratories, in close spatial proximity.[101] Most laboratories focus on one or a few molecules, tracing their interactions and roles in exquisite detail. These results cover only narrow aspects of stem cell phenomena. More comprehensive explanations emerge at higher sociological levels than individual laboratories. One idea, which is still rather preliminary, is that explanations of stem cell phenomena are constructed by *experimenting communities* rather than by individual scientists or laboratories (Fagan 2007, 2013a). If this is correct, then stem cell experiments should be conceptualized not as isolated activities of a single laboratory but as modes of participation in a wider experimenting community. But this is only one suggestion; there are many other ideas to explore.

Conclusion

The main achievements of stem cell biology (so far, anyway) are not theories but striking experimental productions: "immortal" cell lines with unlimited developmental abilities (Thomson et al 1998); embryo-like cells from "reprogrammed" adult cells (Takahashi and Yamanaka 2006); and muscle, blood, and nerve tissue generated from stem cells in culture (Lanza and Atala 2013, and references therein). We currently have no means other than experiments to identify and characterize stem cells – the field has markedly few predictive theories. But this does not mean that philosophers cannot inquire into the ways that we gain scientific knowledge about stem cells. It just means that to do so, philosophers need to look more closely at concrete experimental methods and results, including stem cell lines and their derivatives. New stem cell

---

[100] Regeneration is often conceived as an aspect of development, so the large swathes of stem cell research aimed at understanding and harnessing stem cells' regenerative powers are also encompassed by this framing. See Section 4 for more on stem cells in regenerative medicine.

[101] So the extreme context dependence of stem cells as biological entities is reflected and reinforced by geographically structured variation in laboratory methods and results.

phenomena are continually being created, as technologies are applied to biological materials in new ways. The field is thus open-ended and continually in flux. Experimental systems and models multiply and diversify, generating new phenomena. In this respect, stem cell biology resembles its subject matter: cells differentiating in local environments. This follows from what I have termed "an uncertainty principle for stem cells." Experiments on a single stem cell are necessarily retrospective; stem cell researchers literally "do not know what they've got 'til it's gone."[102] In practice, stem cells are grounded in biological reality in ways inextricably connected to particular experimental contexts. There is no absolute or generally individuated stem cell type, as there is for, say, neurons or red blood cells. The stem cell concept does not map onto one tidy piece of biological reality, but many – each variant relative to an experimental context. Stem cell research is a patchwork of experiments, producing a fabric of models.

4 Using Stem Cells

Alongside scientific knowledge, stem cell research has produced a fantastic variety of cell types, cell lines, and engineered model systems – the concrete products of stem cell experiments. How are these products used? Medical uses predominate. Indeed, stem cell biology is currently unified as a field by the goal of using stem cells to cure or alleviate a wide variety of human pathologies. This section first summarizes the main current uses for stem cells and then turns to uses hoped-for but still unrealized. The gap between the two is stark. I next discuss why stem cells have not (yet) delivered on their apparent medical promise. The last section points to emerging connections between the philosophy of stem cells and other topics in the philosophy of science.

Useful Cells

The raison d'être of stem cell biology is the hope of using (some varieties of) stem cells to treat or cure diseases for which no adequate therapy exists. Without this unifying therapeutic goal, stem cell research would be fragmented into different fields of life and medical science. Even today, as noted, the field is highly dispersed in its concepts and experimental practices – to the point that unifying models are philosophers' constructs rather than tools of stem cell research itself. Such unification is provided by the overarching goal of harnessing stem cell capacities to innovate new therapies.[103] So here is another way stem cell biology differs from the areas of science philosophers tend to study.

---

[102] Apologies to Joni Mitchell.

[103] The field coalesced around this goal in the late 1990s, assuming its present configuration.

Stem cell research is not "pure science" aimed at knowledge for its own sake, but instead seeks "useful knowledge" to improve human health.[104]

There are three main such uses for stem cells: drug development, disease models, and regenerative medicine. My discussion of the first two will be brief. In drug development, human stem cells are a species-specific resource to efficiently test candidate drugs for safety and efficacy.[105] Both stem cell capacities are important to this role: we can get a lot of human cells quickly (via self-renewal) and differentiate them as needed to one or more key cell types. This use adds stem cells to existing biopharmaceutical methods to speed up the path from "bench to bedside"; they are incorporated into methods of clinical translation. A second use is as models of human disease, in efforts to learn about mechanisms underlying a specific pathology. More precisely, stem cell lines and/or their products (including organoids and embryo-like structures) are used as models for learning about molecular mechanisms underlying inherited diseases. This use is closely related to (and involved with) stem cells' role as models of normal human develop-ment, as discussed in Section 3. Induced pluripotent stem cells are particu-larly significant in this regard, offering the prospect of patient-specific stem cell lines for Parkinson's, ALS, muscle wasting diseases, diabetes, and more.

Stem cells' third use is central for regenerative medicine. Instead of adding stem cells to existing clinical platforms or modeling approaches, regenera-tive medicine offers a new therapeutic approach: replace nonfunctioning or diseased cells and tissues with healthy human counterparts. This is stem cells' most novel and distinctive use, so I will discuss it in more detail. Indeed, a medically oriented philosophical account of stem cells might characterize them as essentially regenerative, locating the field's conceptual foundations in regeneration biology rather than studies of organismal development.[106] The therapeutic promise of stem cell research and, accord-ingly, a large share of the public interest in stem cells is based on their regenerative potential. The scope of hoped-for regenerative therapies is

---

[104] This is not to say that all stem cell research projects are clinically oriented. Rather, the therapeutic aim gathers many diverse projects under the same broad umbrella. Stem cell research is not unusual in this regard. Philosophers' own disciplinary inclinations tend to underemphasize the uses and effects of science in society more broadly.

[105] For the same reason, stem cells can also be used in toxicity tests. The role of stem cells in clinical research/evidence-based medicine is, to my knowledge, an unexplored topic in phil-osophy of science.

[106] Thanks to an anonymous reviewer for raising this point. The conceptual anatomy of stem cell research that I present here is not the only possible one. I have chosen to focus on organismal development as the phenomenon anchoring the conceptual core of stem cell research, rather than biological studies of regeneration, because the former strikes me as primary in the sense

very broad. In principle, any disorder or injury involving any cell type or tissue in the body could be ameliorated by targeted manipulation of stem cells or their products. In practice, the primary clinical targets are First World diseases: cancer, diabetes, heart disease, muscular dystrophy, and neurodegenerative conditions such as Parkinson's and Alzheimer's.

These targets of hoped-for stem cell applications have stayed fairly constant for the past decade. Unfortunately, so have realized applications. A 2019 review by De Luca and colleagues identifies three well-confirmed regenerative uses of stem cells:

(1) bone marrow (HSC) transplantation for blood disorders, including but not limited to leukemia;[107]

(2) epithelial stem cells[108] for grafts to treat burns (skin and cornea); and

(3) retinal regeneration to cure blindness due to macular degeneration.

Of these, the first is the most established, an exemplar for much clinically oriented stem cell research. Bone marrow transplantation (BMT) was pioneered in the 1950s, and from its first rather tenuous claims of clinical success, it has evolved into a powerful and effective therapy (Mayo Clinic 2021). The strategy is simple: replace diseased or damaged bone marrow with HSC, much as in "radiation rescue" for mice, discussed earlier. Section 2's abstract model neatly captures the main steps. Bone marrow is taken from an adult human donor, the patient, or another immunologically matched person (Condition 1). A standard set of cell traits is used to (defeasibly and provisionally) isolate HSC from the complex mélange of bone marrow cells (Condition 2). Donor HSC are injected into the patient, where they self-renew and differentiate to produce a healthy blood and immune system (Conditions 3 and 4). Often, the patient's original blood or immune system is removed by radiation or chemotherapy (eliminating, e.g., cancerous blood cells). In any case, a new blood and immune system develop from transplanted donor HSC. More than thirty thousand HSC

---

that regeneration is included as a section of organismal development. (After all, organs and tissues must develop in the first place, before there is *re*-generation.) This conceptual ordering is reflected in, for example, the new ISSCR education guidelines for advanced undergraduates and early graduate students (2020), which proposes a standard course for learning about stem cells beginning with the basics of organismal development, with regeneration introduced halfway through the course. But that is not to say that one could not analyze stem cells through the lens of regeneration, and it would be interesting to compare and contrast such an approach with my account here.

[107] This is sometimes partnered with gene therapy; genome-editing applications are likely on the horizon.

[108] These were first cultured in 1975 but not called "stem cells" at the time. Various clonally growing epidermal cells are classified as stem cells today.

transplants are performed worldwide each year to treat a wide variety of blood disorders.

BMT uses stem cells (HSC) directly, as renewable sources of healthy blood and immune cells. A more indirect regenerative use is to direct mammalian cells to develop along pathways of our choosing, to repair damaged organs and tissues in situ. Rather than deliver stem cells themselves to patients, one applies knowledge of molecular mechanisms underlying stem cell capacities to manipulate human cells' regenerative abilities. Such indirect use of stem cells is sometimes included in BMT. For example, HSC can be found in peripheral (circulating) or umbilical cord blood as well as bone marrow, and in some cases, it is medically preferable to extract HSC from peripheral blood. In such cases, before the transplant, the donor takes growth factors to increase their HSC production and circulation. This puts larger numbers of HSC into the peripheral blood for collection. The basic idea, using a single molecule (drug) to adjust the degree, location, or pattern of stem cell self-renewal or differentiation in a patient's body, has extremely broad application. But using it successfully requires a lot of background knowledge about how those stem cells work in the body. In both direct and indirect aspects, BMT is an exceptional, exemplary use of stem cells. Despite decades of research, we have no others like it.

The other two evidentially supported clinical uses of stem cells are not in routine clinical use but considered experimental (though promising). These uses are limited, interestingly, to localized patches on the surface of human organisms: the skin and eye. Several different stem cell varieties collaborate to form and maintain human skin, and these pathways have been characterized in great detail using mouse models and in vitro human tissue culture.[109] This knowledge, along with advances in tissue engineering, has been used to make skin grafts to treat serious burns. Stem cell–based skin grafts for burns were innovated in the 1980s; extension of this approach to the eye is more recent. This use is somewhat indirect; the appropriate stem cells help make the grafts. More direct use of stem cells to accelerate healing in burns and other wounds is still preclinical and experimental (reviewed in Ahmadi et al 2019). Similarly, retinal regeneration using pluripotent stem cells shows encouraging early results in treating macular degeneration (De Luca et al 2019). But this is not yet an established, routine therapy. BMT is the only evidentially confirmed stem cell treatment in routine clinical use. More than two decades after the field of stem cell biology was reconfigured and vitalized around their regenerative promise, clinical advances using stem cells are very modest.

---

[109] Epithelial stem cells, including those of skin and eye, deserve more study in their own right, though constraints of space do not allow for that here.

Hoped-for applications of stem cells are, in contrast, broad and sweeping. Muscular dystrophy and other muscle-wasting diseases have long been a focus, the idea being that skeletal muscle stem cells (identified circa 1961) could renew that tissue. But despite many attempts to spur muscle regeneration in situ, this has not worked. Cardiac muscle has much less innate regenerative ability than skeletal muscle; twenty years of attempts to coax bone marrow-derived stem cells into this tissue recently imploded with the discovery of data falsification going back to (plausibly) 2001. So-called mesenchymal stem cells have been a locus of misunderstanding and controversy, inspiring numerous clinical trials with dubious scientific rationale.[110] There is considerable excitement about uses for stem cells to treat neurological diseases. For Parkinson's, the rationale is clear: symptoms arise from low dopamine levels, which can be restored (in some patients) by transplanting fetal neural cells. As donations of this tissue are rare, pluripotent stem cells are proposed as an alternative source of cells for transplants. Treatments for diabetes are similarly well-grounded. In summary, there are many ongoing projects aimed at stem cell therapies, some well-grounded and others less so. Yet, so far, stem cells' medical promises remain largely unfulfilled.

Into the gap between expectations and reality have slithered so-called stem cell clinics, which market unproven products directly to the public. In the past decade, these businesses have dramatically expanded in the US and Canada. (As of 2016, there were 351 "direct-to-consumer" stem cell businesses operating in the United States at 570 clinics.[111]) Often, they use the term "stem cell" loosely, referring to preparations of the patient's own fat or bone marrow that may or may not contain stem cells in the sense discussed in this Element (and among scientists). Other products touted as stem cells are preparations from another person's amniotic, placental, or umbilical cord material.[112] Platelet-rich plasma preparations from blood are often lumped in with "stem cells," as are the often-misleadingly named "mesenchymal stem cells." The provenance of marketed cells is sometimes very unclear, though almost without exception they are advertised as stem cells from adult humans only – no embryo use. The main problem with these businesses is not their terminological laxity, however, but

---

[110] This is not to say that all research on mesenchymal stem cells lacks scientific and clinical value. But the uncertainties of finding and characterizing this so-called variety of stem cell are compounded by intense terminological and methodological variability, making the evidential challenges discussed earlier even more severe.

[111] Turner and Knoepfler (2016). A decade ago, most such businesses were outside the US (hence critiques of "stem cell tourism"), but with little pushback from the FDA, these businesses have expanded dramatically within the US.

[112] The clinic I visited in Salt Lake City was of the latter sort, with cheery assurances that no babies were inconvenienced in acquiring the stem cell elixir.

lack of robust evidence for the treatments they offer patients. Typically, these businesses claim that their "stem cells" can benefit patients suffering from a very wide range of symptoms and diseases, from joint pain to Alzheimer's. But, as we have seen, very few stem cell–based therapies have been shown to be effective by scientific or clinical standards. Direct-to-consumer stem cell clinics do not let the absence of evidence deter them, using personal testimonials instead. Vulnerable people, some suffering greatly, are induced to pay out of pocket for "stem cell" treatments. (The clinic whose public information session I attended in 2014 charged about $5,000 per injection, with clients often encouraged to repeat the procedure several times.) "Stem cell" clinics are a facet of broader epistemic pathologies at work in the US culture (and elsewhere). Plausibly, these clinics' combination of overpromising and eschewal of scientific standards will undermine trust in stem cell science more broadly. In this sense, these business ventures are not just independent of stem cell science but antithetical to it. Anti-science clinics thrive in part because well-evidenced stem cell cures have largely failed to materialize. Why?

Why So Slow?[113]

A decade after hESC were first cultured, proponents of stem cell research anticipated "an impending revolution" both in medicine and in our ideas about cell development (e.g., Trounson 2009, xix). There have been profound changes to the latter. But the medical revolution is slow in coming. Perhaps biological reality is such that stem cell therapies can never succeed, barring a few exceptional cases. However, that degree of pessimism is currently unwarranted. We do not yet know enough about stem cells to give up therapeutic hope entirely. Another hasty answer is to blame bioethical controversy. Without political opposition to research using human embryos, the thinking goes, we would already have stem cell cures. But that is too sweeping. Though barriers to funding hESC research have not helped stem cell biology, the scientific side has progressed even so. A more incisive answer, I think, is that clinical translation – the move "from bench to bedside" – presents distinctive challenges for stem cell research. For one thing, what we want to do with stem cells *medically* is exactly what we cannot do with them *experimentally*. The primary therapeutic aim is to use stem cells to repair human bodies; to transplant these powerful regenerative agents into human patients, in whose bodies they grow and maintain healthy organs and tissues.[114] That's

---

[113] Title from Valian (1997).

[114] Or, more indirectly, manipulate human cells' regenerative capacities in situ. The knowledge base needed to do this safely and effectively is much the same as for direct transplant therapy.

the basis of HSC/bone marrow transplantation therapy and the preclinical successes so far. But ethical restrictions prohibit the corresponding experimental strategy: one cannot study stem cells by removing vital organs or tissues from humans and seeing if they grow back after a transplant. So, stem cell research has to approach its medical target indirectly.

Compounding this challenge are the main features of stem cell science discussed in previous sections: experimental relativity, proliferation of concrete models of human development, context dependence, and transformation. Stem cells are diverse and elusive, their identity tied to experimental contexts. The path to clinical translation is a stepwise procedure designed primarily to test chemical entities – drugs – for safety and efficacy. Treating cells as (very complex) drugs requires them to be "a clearly defined class of intrinsically stable biological objects that can be isolated and purified" (Brown et al. 2006, 339). As previous sections have shown at length, stem cells are not like this – certainly not as a general category, and not even for major varieties such as HSC, ESC, or iPSC. Plausibly, the process of clinical translation needs to be re-calibrated to admit therapeutic agents that do not behave like biochemical species. The latter are robust, specific, and stable – and these characteristics are entrenched in preclinical and clinical requirements. Stem cells, however, are notoriously variable, dynamic, and transformative. So new proposals for standardization and control, better-suited to stem cells identified by experiment, seem called for. Philosophers of science and medicine could contribute to that project, which is recognized by many stem cell researchers as crucial for progress in regenerative medicine.

Of course, to safely and effectively regenerate human bodies using stem cells, much more is needed than new ideas about standardization and new conceptions of control. It is no accident that the lone "systemic," established stem cell therapy is blood-based. Blood is a uniquely separable, free-flowing tissue that can be removed and reinserted from a living human body. Other organs and tissues do their work in situ, where contextual factors (e.g., cell niches) play more fine-grained and variegated roles. To safely regenerate bodily organs and tissues, one needs to know, at a minimum: (1) the optimal cell type(s) to transplant, (2) how to deliver transplanted cells to tissues, (3) how to forestall transplant rejection and other undesirable immune responses, and (4) how to prevent ensuing tumors (the connections between stem cells and cancer run deep). There is no short-cut for the hard biological and biomedical work of gaining this knowledge, for any regenerative use of stem cells in humans. So a key question is: Which lines of work are best-suited to bridge the gap between stem cell promises and clinical practice?

Developmental intermediates like organoids (see Section 3) are one promising approach. As noted, blood is a unique tissue in that there is no three-dimensional spatial structure to manage in designing regenerative therapies. Blood and immune cells circulate in the body and among lymphoid organs. Just get enough HSC into a patient's body, and those stem cells find their own place and do the rest. But for other tissues and organs, regeneration needs to work in a (relatively) stable spatial arrangement – a complex interactive niche. A good way to gain the knowledge needed for safe and effective stem cell–based regenerative therapies is to build these niches in vitro and experimentally discover ways of manipulating human cells' regenerative capacities in these organ-like environments. Organoids, as discussed earlier, are simplified three-dimensional model organs derived from stem cells, which consist of multiple organ-specific cell types in relative positions similar (in some respect) to the mature organ, exhibiting at least one organ-specific function. These similarities in composition, function, and spatial arrangement are taken as evidence that the developmental process that produces an organoid is similar to in vivo development (which, in humans, cannot be observed). In this way, organoids are surrogate research objects, standing in for inaccessible processes of human organ development. This modeling role easily extends to the repair of damaged or diseased tissue. Moreover, a new frontier in organoid research is to increase their complexity and thereby resemblance to human organs – even whole organisms. There are already plans to model whole-organism homeostasis by combining multi-organoid systems into a single integrated model, as well as more advanced organoids containing blood vessels and/or neural connections allowing for greater tissue complexity and responsiveness.

This important strand of organoid research aims at increasingly better approximations of the real thing, perhaps even human organs for transplant.[115] Interconnected combinations of organoids, and models of increasingly complex tissue assemblages, are already works-in-progress for tissue bioengineers. They might also be the leading edges for clinical translation of stem cell research.

There are other aspects of stem cell clinical translation which could benefit from philosophers' attention. As discussed, stem cell research is highly disunified, proceeding via a continuously proliferating fabric of models that reveals aspects of (human) development, piecemeal and fragmentary. This social epistemic organization may itself be an obstacle to clinical translation. What forms of organization would most effectively facilitate it? Scientists have suggested instituting interdisciplinary centers

---

[115] See Fagan (2020) for further discussion.

for "stem cell excellence" that encourage cross-disciplinary collaboration between clinicians, veterinarians, basic researchers, bioinformatics experts, tissue engineers, visualization specialists, and more. Relatedly, there are calls to integrate and standardize data about "curated" stem cell lines and their products, clinical information, and genomics (e.g., Theo Murphy 2015, 9; Rackham et al 2021). Standardization of stem cell lines (i.e., concrete models themselves) is another rallying-cry. There are many opportunities here for social epistemologists (including those using formal models) to explore possibilities and consider alternative ways the field of stem cell research might be organized.

Another approach would connect these issues with feminist philosophy and other social justice ideas. There are a number of feminist studies of stem cell biology and related fields. For example, Waldby and Cooper (2010) discuss women as bioeconomic actors in the "tissue economy" of cord blood, eggs, embryos, examining the implications of this aspect of the field on women's opportunities and autonomy. Reflections on stem cell-derived embryo models could be useful for practices of assisted reproduction, infertility treatment, and contraception. It would be valuable to extend bioethics research on stem cells in the direction of social justice. A different but compatible tack would be an analysis of the spectacular failure of cardiac muscle stem cell research (Murry and MacLellan 2020). To sketch the problem very briefly: the American Heart Association does not fund hESC research. This created an incentive for researchers to find "adult stem cells" (i.e., not derived from human embryos) that can regenerate cardiac muscle: HSC, mesenchymal stem cells, and muscle progenitors. Clinical and preclinical studies pursued this idea from the 1990s onward, with mixed results. Recently, the positive studies were found to be based on fraudulent work, leading to the retraction of more than thirty papers. The actual results, when fabrications were removed from the record, showed little regenerative benefit from the stem cell types that had been tried.[116] For nearly twenty years, cardiac stem cell researchers had been pursuing a politically expedient phantom. What forms of social organization could prevent similar incidents?

The challenge to clinical translation of experimenters' knowledge of stem cells can also be approached in terms of scientific models and modeling. The ultimate goal of stem cell research is to provide replacement parts for human bodies or to coax our bodies into making those parts for ourselves. The

---

[116] "Although each of these adult cell types was originally postulated to differentiate directly into cardiomyocytes, none of them actually do" (Murry and MacLellan 2020, 854). Researchers are now using pluripotent stem cell lines (iPSC and ESC).

results of stem cell research are intended for use in the body – they are intended to become *us*. This erases the distinction between concrete stem cell models and their primary target: cells developing and functioning in human bodies. That is, regenerative use of stem cells in vivo requires eliding the model-target relation that frames much of stem cell biology. No wonder the process has not been speedy! Stem cells in vitro are not the same as those in our bodies; in some ways, they are constructed not to be so. Models are not identical to their targets. What is being asked of clinical translation of stem cells for regenerative purposes is, perhaps, in conflict or marks a tradeoff with, the modeling role of stem cells in scientific practice. Working out ways to reconcile the modeling and therapeutic roles of stem cells and their experimental relatives is another important task for philosophers of science.

New Directions

So there are multiple ways philosophers of science could help facilitate the clinical translation of stem cells. But there are also other interesting points of contact between applications of stem cell biology and philosophy of science. In the space remaining, I discuss a few. (Because bioethics debates pertaining to stem cell research are familiar, well-trodden ground, I do not include those here.)

Combining Experiments

Converging lines of independent evidence play an important role in confirming scientific hypotheses and theories. What about converging experimental methods? This is a common pattern in stem cell research: stem cell experiments frequently combine different biotechnologies into a more inclusive, elaborate experimental assemblage. For example, stem cell techniques frequently pair with CRISPR/Cas genome editing: "precise manipulation of cellular DNA sequences to alter cell fates and organism traits" (Doudna 2020).[117] Targeted genome editing, of course, has an enormous range of potential applications: scientific, industrial, medical, and environmental. Stem cells are a partner in many of these applications. One important approach is to eliminate mutations that cause disease – edit the genome

---

[117] "CRISPR" is an acronym for "clustered regularly interspaced short palindromic repeat" – literally, a section of a genome exhibiting clusters of short palindromic DNA base sequence repeats. These sequence motifs are found in micro-organisms, where, coupled with enzymes, they play a role in immune defense. Jennifer Doudna, Emmanuelle Charpentier, and their colleagues characterized and adapted this micro-organismal immune mechanism to edit other organisms' genomes in a targeted way (Jinek et al. 2012).

to remove or correct the "mistake" that leads to an organism's disease in an organism. For genetic diseases of blood cells, such as leukemia, HSC are to be vehicles of genetic cures.[118] The rationale is straightforward. One edited cell is likely to make little difference in an organism. But a few edited cells with the capacity to self-renew and differentiate into all the cells of an organ, tissue, or whole organism – those could be effective. Genome editing combines with ESC and iPSC culture methods to generate modified cells of potentially any type – even modified cells specific to a particular patient.[119] Of course, the therapeutic use of edited stem cells faces the same challenges as discussed previously, alongside new safety concerns. Organoids, however, are fair game for genome editing; the combination of iPSC, organoids, and genome editing makes for a powerful scientific tool. Other technologies that increasingly partner with stem cell experiments include single-cell RNA and genome sequencing, bioinformatics tools, immunology (for matching cell and tissue transplants), and machine learning. In brief, stem cell methods play well with other biotechnologies, and their opportunistic, effective combination is a hallmark of stem cell research. The impact of this feature on our knowledge of stem cells and related phenomena also warrants further study – and may be common in experimental fields, decentralized ones, engineering and synthetic sciences, or all of these. There is much here for philosophers of science to pursue.

New Directions in Neuroscience

Another major focus of preclinical stem cell research is the prospect of central nervous system (CNS) cell replacement.[120] Diseases thought to be good candidates for such an approach are pathologies of one or a few closely related cell types localized to one brain region. These include Parkinson's disease, epilepsy, and eye degeneration (the last already well-confirmed). Stroke and spinal injury, though often discussed as possible foci for stem cell therapy, are now thought to be poorer prospects, due to the many cell types involved and extensive extracellular damage. Without a healthy niche,

---

[118] "In principle, sickle cell disease could be cured by removing blood stem cells—that is, haematopoietic progenitor cells—from a patient and using genome editing to either correct the disease-causing mutation in β-globin or activate expression of γ-globin, a fetal form of haemoglobin that could substitute for defective β-globin. . . . The edited stem cells could then be transplanted back into the patient, in whom the progeny of these edited stem cells would produce healthy red blood cells" (Doudna 2020, 230).

[119] iPSC are increasingly suggested as patient-specific resources, "a streamlined strategy for CRISPR therapeutic development" (Doudna 2020, 234).

[120] Or by using stem cells to mobilize endogenous mechanisms, but this is less emphasized in reviews I have seen.

stem cells' regenerative powers seem likely to be thwarted. There is unease in the stem cell community about clinical trials that press forward with dubious biological rationale (e.g., neurological applications of umbilical cord blood or "mesenchymal stem cells"). Using stem cells to model neurological disease has so far been more successful, through the innovation of cerebral organoids – i.e., "mini-brains" or "brains in a dish." Jurgen Knoblich's group in Vienna models Zika virus-related microcephaly using iPSC, genome editing, and cerebral organoids. Researchers at the Rockefeller Institute model Huntington's disease model in "neuraloids" – embryo-like structures grown from CRISPR-modified ESC, which reprise differentiation of cell types involved in ectoderm formation, as well as the formation of structures like the neural tube. Data analysis in this system includes single-cell sequencing and machine learning – a further confluence of technologies.

Perhaps most intriguing for philosophers of neuroscience, psychology, and mind is the recent discovery that neuroloids cultured for six to nine months produce brainwaves – that is, they exhibit electrical patterns of spikes forming waves, "microcircuitries" that indicate cell-cell communication via neural networks. At this stage, these neuraloids are lumpy spheres about the size of a pea, far simpler than a human brain. Yet machine-learning algorithms cannot distinguish the in vitro electrical patterns from fetal ones. Electrically excitable mini-brains have been proposed as disease models for bipolar disorder, schizophrenia, and autism. It has not escaped researchers' notice that, as brain organoids become more complex, and more closely approximate human brains, they might approximate certain kinds of mental activities. There's been concern about preventing organoid consciousness – not in the immediate future, but eventually. Organoids could be a tool for conceptualizing and studying mental function, cognition, and consciousness, extending their modeling role to the psychological sphere. There are, evidently, many interesting issues for philosophers to consider, in stem cell research aimed at CNS repair – not least, relations between biological and psychological domains, concepts, and processes.

Personalized Medicine

iPSC are produced by manipulating gene expression in a specialized cell, to induce an embryo-like state (see Section 3). Because they can in principle be made from just about any cell of a mammal's body, iPSC can be made from particular individuals – and seem tailor-made for personalized medicine. Their proliferative potential, both as lineage-producing cells and as models for individual human patients, is just beginning to be realized. The prospect of editing iPSC from a particular human patient, so as to generate unlimited healthy cells that pose no immunological obstacle to transplantation, is extremely appealing. iPSC, that is, could bypass problems of immune rejection for cell and tissue transplants. In light of this potential, it is interesting that so many scientists and policy-makers opt instead for "democratized medicine" over personalized iPSC lines. The issue is cost. Cell lines are expensive to make and maintain, and so it is thought more practical to create "iPSC banks" of cell lines that use standard organ-matching criteria, as "off-the-shelf cell products that meet broad population needs" (Theo Murphy 2015, 4).[121] (The off-the-shelf/bespoke analogy seems apt in this case.) It seems worth asking, in the broader context of medicine and society, what grounds this decision, and to identify the key assumptions involved. In what ways are iPSC lines representative of a species, an individual person, a tissue compatibility type, a disease, and so forth? What concept of iPSC makes the best use of these proliferative stem cells?

Epilogue

As the previous sections indicate, there is much more to do in philosophy of stem cell biology. The field of stem cell research (like many fast-moving, highly technical experimental sciences) is under-explored terrain for philosophers. This short book offers some conceptual tools and insights for future work. I briefly summarize these, by way of conclusion.

Stem cells are a rich source of insights about organisms and development, as well as about scientific practices and knowledge. To extract them, some conceptual background is needed: ideas of cells, cell type, differentiation, and lineage – all implicated in the process of organismal development. The stem cell concept thus gives philosophers a window on that process, as studied in an experiment-driven and highly technical field. Examining the conceptual background and core of the field reveals that stem cells are not a cell type in the way of neurons, red blood cells, and so forth. They are, instead, defined by what they give rise to rather than the cell traits they exhibit. Section 2 traced the historical origins of this idea, noting contrasts with present-day definitions. The latter converge on two cell processes: self-renewal and differentiation. I further explicate these ideas of cell reproduction and development, using the form of lineage trees and indicating connections with the organism-level. These insights are integrated into an abstract model that clarifies the stem cell concept. Importantly, though abstract, this model links directly to stem cell experiments, in that specific varieties of stem cell that

---

[121] Proposals to use iPSC in chimeric antigen receptor therapy (CAR T) for leukemia raise this same issue.

figure in stem cell research correspond to specifications of values of the variables in the model. In scientific practice, these specifications are made by materials and methods of experiments, with wide variation across laboratories and research projects within the field. So the model reveals how the stem cell concept, in scientific practice, is entangled with experimental methods for identifying and characterizing stem cells.

Section 3 looked deeper into those methods, using three exemplars to characterize their basic experimental design. Building on the conceptual insights of Section 2, I showed that stem cell varieties are identified relative to their experimental contexts and that identifying a single cell as a stem cell of any particular variety is unavoidably uncertain (an "uncertainty principle" for stem cells). This, in turn, motivates what I term the "single-cell standard" for stem cell research: scientists seek to meet the evidential challenge posed by stem cell uncertainty, and new experimental techniques that bring the field closer to this ideal are greeted as important successes. In this way, the abstract model and analysis of core stem cell experiments offer a robust framework with which to philosophically examine an otherwise dauntingly heterogeneous, technical, decentralized, and fast-moving field. Relatedly, I have argued that knowledge in stem cell biology takes the form of a patchwork of experiments, producing a fabric of models. Section 4 considered the main uses for stem cells, realized and hoped-for, and proposed reasons for the longstanding gap between the two. Perhaps most significantly, I've indicated potential future directions for philosophical studies of stem cells and stem cell research. Accounts of biological individuality, organismal development, the cell-organism relation, models and modeling, clinical translation, experiment – stem cell research bears on all these and more. As that field continues to weave its fabric of experiments and models, even more connections should appear. Stem cells' own generative capacities reflect their epistemic potential, for philosophers as well as scientists.

References

Ahmadi, AR, Chicco, M, Huang, J, Qi, L, Burdick, J, Williams, GM, Cameron, AM, and Sun, Z (2019). Stem cells in burn wound healing: A systematic review of the literature. *Burn* 45: 1014–1023.

Andrews, P (2002). From teratocarcinomas to embryonic stem cells. *Philosophical Transactions of the Royal Society of London, Series B* 357: 405–417.

Blasimme, A, Schmietow, B, and Testa, G (2013). Reprogramming potentiality: The co-production of stem cell policy and democracy. *American Journal of Bioethics* 13: 30–32.

Borgés, J-L (1998). On exactitude in science. In, Jorge Luis Borges, *Collected Fictions* (Trans. Hurley, H.) New York: Penguin Books, 325.

Boveri, T. (1892). Über die Entstehung des Gegensatzes zwischen den Geschlechtszellen und die somatischen Zellen bei Ascaris megalocephela. *Sitzungsberichte der Gesellschaft für Morphologie und Physiologie in München* 8: 114–125.

Brandt, C (2012). Stem cells, reversibility, and reprogramming. In Mazzolini, R, and Rheinberger, H-J (eds.), *Differing Routes to Stem Cell Research: Germany and Italy*. Bologna: Società editrice il Mulino, 55–91.

Brown, N, Kraft, A, and Martin, P (2006). The promissory pasts of blood stem cells. *BioSocieties* 1: 329–348.

Brumbaugh, J, Di Stefano, B, and Hochedlinger, K (2019). Reprogramming: Identifying the mechanisms that safeguard cell identity. *Development* 146: dev182170.

Bursten, J (ed.) (2019). *Perspectives on Classification in Synthetic Sciences*. Routledge, London and New York.

Can, A (2008). A concise review on the classification and nomenclature of stem cells. *Turkish Journal of Hematology* 25: 57–59.

Cell Therapy and Regenerative Medicine Glossary (2012). Stem cell. *Regenerative Medicine*, 7, S1–S124.

Chang, H (2004). *Inventing Temperature*. Oxford: Oxford University Press.

Coleman, W (1977). *Biology in the Nineteenth Century: Problems of Form, Function, and Transformation*. Cambridge: Cambridge University Press.

Cooper, M (2003). Rediscovering the immortal Hydra. *Configurations* 11: 1–26.

Crisan, M, and Dzierzak, E (2016). The many faces of hematopoietic stem cell heterogeneity. *Development* 143: 4571–4581.

De Luca, M, Aiuti, A, Cossu, G, Parmar, M, Pellegrini, G, and Robey, PG (2019). Advances in stem cell research and therapeutic development. *Nature Cell Biology* 21: 801–811.

Downes, S (1992) The importance of models in theorizing: a deflationary semantic view. *Proceedings of the Biennial Meeting of the Philosophy of Science Association*, Vol. 1992, Volume One: Contributed Papers, 142–153

Dröscher, A (2002). Edmund B. Wilson's "The Cell" and cell theory between 1896 and 1925. *History and Philosophy of the Life Sciences*, 24: 357–389.

Dröscher, A (2012). Where does stem cell research stem from? In Mazzolini, R, and Rheinberger, HJ (eds.), *Differing Routes to Stem Cell Research: Germany and Italy*. Bologna: Società editrice il Mulino, 19–54.

Dröscher, A (2014). Images of cell trees, cell lines, and cell fates: The legacy of Ernst Haeckel and August Weismann in stem cell research. *History and Philosophy of the Life Sciences* 36: 157–186.

Doudna, J (2020). The promise and challenge of therapeutic genome editing. *Nature* 578: 229–236.

Dupré, J, and Nicholson, DJ (2018). A manifesto for a processual philosophy of biology. In Nicholson, DJ, and Dupré, J (eds.), *Everything Flows*. Oxford: Oxford University Press 3–45.

European Stem Cell Network (2016). Stem cell glossary. Available at www.eurostemcell.org/stem-cell-glossary#letters. Accessed March 13, 2016.

Fagan, MB (2007). The search for the hematopoietic stem cell: Social interaction and epistemic success in immunology. *Studies in History and Philosophy of Biological and Biomedical Sciences* 38: 217–237.

Fagan, MB (2010). Social construction revisited. *Philosophy of Science* 77: 92–116.

Fagan, MB (2011). Social experiments in stem cell biology. *Perspectives on Science* 19: 235–262.

Fagan, MB (2013a). *Philosophy of Stem Cell Biology*. London: Palgrave Macmillan.

Fagan, MB (2013b). The stem cell uncertainty principle. *Philosophy of Science* 80: 945–957.

Fagan, MB (2015a). Crucial stem cell experiments? Stem cells, uncertainty, and single-cell experiments. *Theoria* 30: 183–205 (Special Section: Philosophy of Experiment).

Fagan, MB (2015b). Explanatory interdependence: the case of stem cell reprogramming. In Braillard, Pierre-Alain, and Malaterre, Christophe (eds.), *Explanation in Biology: An Enquiry into the Diversity of Explanatory Patterns in the Life Sciences*. Dordrecht: Springer, 387–412.

Fagan, MB (2016a). Cell and body: Individuals in stem cell biology. In Guay, A, and Pradeu, T (eds.), *Individuals across the Sciences*. Oxford: Oxford University Press, 122–143.

Fagan, MB (2016b). Generative models: Human embryonic stem cells and multiple modeling relations. *Studies in History and Philosophy of Science, Part A*, 56: 122–134.

Fagan MB (2017). Stem cell lineages: Between cell and organism. *Philosophy and Theory in Biology* 9, Special Issue: Ontologies of Living Beings. www .philosophyandtheoryinbiology.org.

Fagan, MB (2018). Individuality, organisms, and cell differentiation. In Bueno, Otávio, Chen, Ruey-Lin, and Bonnie Fagan, Melinda (eds.), *Individuation across Experimental and Theoretical Sciences*. Oxford: Oxford University Press, 114–136.

Fagan, MB (2019). Stem cell lineages and classification. In Bursten, J. (ed.), *Perspectives on Classification in Synthetic Sciences: Unnatural Kinds*. New York: Taylor and Francis, 114–135.

Fagan, MB (2020). Organoids: A vital thread in a generative fabric of models. Published in German (trans. Anja Pichl), in *Organoide: Ihre Bedeutung für Forschung, Medizin und Gesellschaft* (*Organoids: Their Importance for Research, Medicine and Society*), edited by Bartfeld, S, Schickl, H, Alev, C, Koo, B-K, Pichl, A, Osterheider, A, and Marx-Stölting, L. Baden-Baden: Nomos, 149–170

Franklin, S (2013). *Biological Relatives: IVF, Stem Cells, and the Future of Kinship*. Durham: Duke University Press.

Giere, RN (2010). An agent-based conception of models and scientific representation. *Synthese* 172: 269–281.

Godfrey-Smith, P (2006). The strategy of model-based science. *Biology and Philosophy* 21: 725–740.

Guay, A, and Pradeu, T (eds.) (2016). *Individuals across the Sciences*. Oxford: Oxford University Press.

Hacking, I. (1983). *Representing and Intervening*. Cambridge: Cambridge University Press.

Haeckel, E (1876). *The History of Creation*, translation revised by E. Ray Lancaster (Vol. 2). New York: D. Appleton and Co.

Haeckel, E (1905). *The Evolution of Man*, translated from the 5th (enlarged) edition by Joseph McCabe (Vol. 2). New York: G. P. Putnam's Sons.

Harris, H (2000). *The Birth of the Cell*. New Haven: Yale University Press.

Hauskeller, C, Manzeschke, A, and Pichl, A (eds.) (2019). *The Matrix of Stem Cell Research: An Approach to Thinking of Science and Society*. London: Routledge.

Hooke, R (1665). *Micrographia*. London: The Royal Society of London.

Hopwood, N (2005). Visual standards and disciplinary change: Normal plates, tables and stages in embryology. *History of Science*, 43: 239–303.

Hyun, I, Munsie, M, Pera, MF, Rivron, NC, and Rossant, J (2020). Toward guidelines for research on human embryo models formed from stem cells. *Stem Cell Reports* 14: 169–174.

International Society for Stem Cell Research (2016). Stem Cell Glossary. Available at www.isscr.org/visitor-types/public/stem-cell-glossary#stem . Accessed June 8, 2017.

International Society for Stem Cell Research (2020). Core Concepts in Stem Cell Biology: Syllabus and Learning Guide. Available for download at www .isscr.org.

James, W (1890). *The Principles of Psychology*. Cambridge, MA: Harvard University Press.

Jinek, M, Chylinski, K, Fonfara, I, Hauer, M, Doudna, JA, and Charpentier, E (2012). A programmable dual-RNA-guided DNA endonuclease in adaptive bacterial immunity. *Science* 337: 816–821.

Keating, P, and Cambrosio, A (2003). *Biomedical Platforms*. Cambridge: The MIT Press.

Kendig, C (2016). Activities of Kinding in Scientific Practice. In Kendig C (ed.), *Natural Kinds and Classification in Scientific Practice*. London: Routledge, 1–13.

Kobold, S, Guhr, A, Kurtz, A, and Löser, P (2015). Human embryonic and induced pluripotent stem cell research trends: Complementation and diversification in the field. *Stem Cell Reports* 4: 914–925.

Kraft, A (2009). Manhattan transfer: Lethal radiation, bone marrow transplantation, and the birth of stem cell biology, 1942–1961. *Historical Studies in the Natural Sciences* 39: 171–218.

Kurtz, A, Seltmann, S, Bairoch, A, Bittner, M-S, Bruce, K, et al. (2018). A standard nomenclature for referencing and authentication of pluripotent stem cells. *Stem Cell Reports* 10: 1–6.

Lancaster, MA, and Knoblich, JA (2014). Organogenesis in a dish: Modeling development and disease using organoid technologies. *Science* 345: 1247125-1-8.

Landecker, H (2007). *Culturing Life. How Cells Became Technologies*. Cambridge, MA: Harvard University Press.

Lanza, R, Gearhart, J, Hogan, B, Melton, D, Pederson, R, Thomas, E, Thomson, J, and Wilmut, I (2009). (eds.) *Essentials of Stem Cell Biology*, 2nd edition. San Diego, CA: Academic Press.

Lanza R and Atala A (2013) (eds.). *Essentials of Stem Biology*, 3rd ed. San Diego: Academic Press.

Liu, D (2018). Heads and tails. In Matlin, K, Maienschein, J, and Laubichler, M (eds.). *Visions of Cell Biology*. Chicago: University of Chicago Press, 209–245.

Maehle, AH (2011). Ambiguous cells. *Notes and Records of the Royal Society* 65: 359–378.

Maherali, N, and Hochedlinger, K (2008). Guidelines and techniques for the generation of induced pluripotent stem cells. *Cell Stem Cell* 3: 595–605.

Maienschein, J (2003). *Whose View of Life?* Cambridge, MA: Harvard University Press.

Martin, G (1981). Isolation of a pluripotent stem cell line from early mouse embryos cultured in medium conditioned by teratocarcinoma stem cells. *Proceedings of the National Academy of Sciences, US* 78: 7634–7638.

Matlin, KS, Maienschein, J, and Laubichler, MD (eds.) (2018). *Visions of Cell Biology*. Chicago: University of Chicago Press.

Maximow, A (1909) Der Lymphozyt als gemeinsame Stammzelle der verschiedenen Blutelemente in der embryonalen Entwicklung und im postfetalen Leben der Säugetiere. *Folia Haematologica* 8: 125–141

Mayo Clinic (2021). Patient care and health information: Bone marrow transplant. Mayo Foundation for Medical Education and Research, www.mayoclinic.org/tests-procedures/bone-marrow-transplant/about/pac-20384854. Accessed January 17, 2021.

Melton, D (2013). Stemness: Definitions, criteria, and standards. In Lanza, R, and Atala, A (eds.), *Essentials of Stem Biology*, 3rd ed. San Diego: Academic Press, 7–17.

Mesa, KR, Rompolas, P, and Greco, V (2015). The dynamic duo: Niche/stem cell interdependency. *Stem Cell Reports* 4: 961–966.

Murry, CE, and MacLellan, WR (2020). Stem cells and the heart – the road ahead. *Science* 367: 854–855.

National Institutes of Health (2016). US Department of Health and Human Services. Glossary. Stem Cell Information. Available at http://stemcells .nih.gov/glossary/Pages/Default.aspx. Accessed March 13, 2016.

Neto, C (2019). What is a lineage? *Philosophy of Science* 86: 1099–1110.

Nishizawa, M, Chonobayashi, K, Nomura, M, Takaori-Kondo, A, Yamanaka, S, and Yoshida, Y (2016). Epigenetic variation between human induced pluripotent stem cell lines is an indicator of differentiation capacity. *Cell Stem Cell* 19: 341–354.

Nobel Foundation (2012). The 2012 Nobel Prize in Physiology or Medicine – Press Release. Nobelprize.org. October 10, 2012, www.nobelprize.org /nobel_prizes/medicine/laureates/2012/press.html.

O'Malley MA (2014). *Philosophy of Microbiology*. Cambridge: Cambridge University Press.

Pennisi, E (2018). Chronicling embryos, cell by cell, gene by gene. *Science* 360: 367.

Potten, CS, and Lajtha, LG (1982). Stem cells versus stem lines. *Annals of the New York Academy of Sciences* 397: 49–61.

Potten, CS, and Loeffler, M (1990) Stem cells: attributes, cycles, spirals, pitfalls and uncertainties. *Development* 110: 1001–1020

Pradeu, T (2012). *The Limits of the Self*. Oxford: Oxford University Press.

Rackham, O, Cahan, P, Mah, N, Morris, S, Ouyang, JF, Plant, AL, Tanaka, Y, and Wells, C (2021). Challenges for computational stem cell biology. *Stem Cell Reports* 16: 3–9.

Ramalho-Santos, M, and Willenbring, H (2007). On the origin of the term "stem cell." *Cell Stem Cell* 1: 35–38.

Rao, M (2004) Stem sense: a proposal for the classification of stem cells. *Stem Cells and Development* 13: 452–455.

Reynolds, AS (2007). The theory of the cell state and the question of cell autonomy in nineteenth and early twentieth century biology. *Science in Context* 20: 71–95.

Reynolds, AS (2018). *The Third Lens*. Chicago: University of Chicago Press.

Schwann, TH (1847). *Microscopical researches into the accordance in the structure and growth of animals and plants*. London: The Sydenham Society.

Science editors (2019). Special section: Approximating organs. *Science* 364: 946–965.

Shamblott, M, Axelman, J, Wang, S, Bugg, E, Littlefield, J, Donovan, P, Blumenthal, P, Huggins, G, and Gearhart, J (1998). Derivation of pluripotent stem cells from cultured human primordial germ cells. *Proceedings of the National Academy of Sciences, US* 95: 13726–13731.

Simian, M, and Bissell, MJ (2017). Organoids: A historical perspective of thinking in three dimensions. *Journal of Cell Biology* 216: 31–40.

Skloot, R (2010). *The Immortal Life of Henrietta Lacks*. New York: Crown Publishing.

Sornberger, J (2011). *Dreams and Due Diligence*. Toronto: University of Toronto Press.

Steinle, H (2002). Experiments in history and philosophy of science. *Perspectives on Science* 10: 408–432.

Stem Cell Reports Q&A (2020). A conversation with John Gurdon and Shinya Yamanaka. *Stem Cell Reports* 14: 351–356.

Takahashi, S, and Yamanaka, S (2006). Induction of pluripotent stem cells from mouse embryonic and adult fibroblast cultures by defined factors. *Cell* 126: 663–676.

Takahashi, K, Tanabe, K, Ohnuki, M, Narita, M, Ichisaka, T, Tomoda, K, and Yamanaka, S (2007). Induction of pluripotent stem cells from adult human fibroblasts by defined factors. *Cell* 131: 861–872.

Theo Murphy High Flyers Think Tank (2015). Recommendations. White paper (Australia).

Thomson, J, Itskovitz-Eldor, J, Shapiro, S, Waknitz, M, Swiergel, J, Marshall, V, and Jones, J (1998). Embryonic stem cell lines derived from human blastocysts. *Science* 282: 1145–1147.

Till, J, and McCulloch, E (1961). A direct measurement of the radiation sensitivity of normal mouse bone marrow cells. *Radiation Research* 14: 213–222.

Trounson, A (2009). Why stem cell research? In Lanza, R, et al. (eds.), *Essentials of Stem Cell Biology*, 2nd ed. San Diego: Academic Press, p. xix.

Turner, L, and Knoepfler, P (2016). Selling stem cells in the USA. *Cell Stem Cell* 19: 1–4.

Valian, V (1997). *Why So Slow? The Advancement of Women*. Cambridge, MA: The MIT Press.

Waldby, C, and Cooper, M (2010). From reproductive work to regenerative labour. *Feminist Theory* 11: 3–22.

Warmflash, A, Sorre, B, Etoc, F, Siggia, E, and Brivanlou, AH (2014). A method to recapitulate early embryonic spatial patterning in human embryonic stem cells. *Nature Methods* 11: 847–854.

Weisberg, M (2013). *Simulation and Similarity*. Oxford: Oxford University Press.

Wilmut, I, Sullivan, G, and Chambers, I (2011). The evolving biology of cell reprogramming. *Philosophical Transactions of the Royal Society B* 366: 2183–2197.

Yin X, Mead BE, Safaee H, Langer R, Karp JM, and Levy O (2016). Engineering stem cell organoids. *Cell Stem Cell* 18: 25–38.

Ying, Q-L, and Smith, A (2017). The art of capturing pluripotency: creating the right culture. *Stem Cell Reports* 8: 1457–1464.

Acknowledgments

Many thanks to Michael Ruse and Grant Ramsey for the invitation to contribute a book to this series. The sections herein have benefited enormously from discussions with my Fall 2018 Philosophy of Biology class (University of Utah), Collin Rice and his Philosophy of Biology class of Fall 2019 (Bryn Mawr), Dick Burian, Julia Bursten, Ubaka Ogbogu, Aleta Quinn, Joe Rouse, and my colleagues at University of Utah, especially Steve Downes, Leslie Francis, Matt Haber, and Anne Peterson. Two anonymous reviewers for Cambridge University Press also provided insightful comments that led to improvements in the final manuscript. Thanks also to audiences at the Center for Philosophy of Science (University of Pittsburgh), San Francisco State University, Virginia Tech, SJ Quinney College of Law (University of Utah), the 2017 International Society for the History, Philosophy, and Social Studies of Biology (São Paulo, Brazil), University of Idaho, participants in the invited session on stem cell clinics at the 2018 APA Pacific Division Meeting, Rice University, University of Cincinnati, and the Boston Colloquium in Philosophy of Science (October 2018). My deepest thanks to members past and present of the Weissman lab at Stanford University Medical Center, who took the time to answer my questions and share their ideas.

Cambridge Elements ≡

Elements in the Philosophy of Biology

Grant Ramsey
KU Leuven

Grant Ramsey is a BOFZAP research professor at the Institute of Philosophy, KU Leuven, Belgium. His work centers on philosophical problems at the foundation of evolutionary biology. He has been awarded the Popper Prize twice for his work in this area. He also publishes in the philosophy of animal behavior, human nature and the moral emotions. He runs the Ramsey Lab (theramseylab.org), a highly collaborative research group focused on issues in the philosophy of the life sciences.

Michael Ruse
Florida State University

Michael Ruse is the Lucyle T. Werkmeister Professor of Philosophy and the Director of the Program in the History and Philosophy of Science at Florida State University. He is Professor Emeritus at the University of Guelph, in Ontario, Canada. He is a former Guggenheim fellow and Gifford lecturer. He is the author or editor of over sixty books, most recently *Darwinism as Religion: What Literature Tells Us about Evolution; On Purpose; The Problem of War: Darwinism, Christianity, and their Battle to Understand Human Conflict;* and *A Meaning to Life.*

About the Series

This Cambridge Elements series provides concise and structured introductions to all of the central topics in the philosophy of biology. Contributors to the series are cutting-edge researchers who offer balanced, comprehensive coverage of multiple perspectives, while also developing new ideas and arguments from a unique viewpoint.

Cambridge Elements ☰

Philosophy of Biology

Elements in the Series

The Biology of Art
Richard A Richards

The Darwinian Revolution
Michael Ruse

Ecological Models
Jay Odenbaugh

Mechanisms in Molecular Biology
Tudor M Baetu

The Role of Mathematics in Evolutionary Theory
Jun Otsuka

Paleoaesthetics and the Practice of Paleontology
Derek D Turner

Philosophy of Immunology
Thomas Pradeu

The Challenge of Evolution to Religion
Johan De Smedt and Helen De Cruz

The Missing Two-Thirds of Evolutionary Theory
Robert N Brandon and Daniel W McShea

Games in the Philosophy of Biology
Cailin O'Connor

How to Study Animal Minds
Kristin Andrews

Inheritance Systems and the Extended Evolutionary Synthesis
Eva Jablonka and Marion J Lamb

Reduction and Mechanism
Alex Rosenberg

Model Organisms
Rachel A Ankeny and Sabina Leonelli

Comparative Thinking in Biology
Adrian Currie

Social Darwinism
Jeffrey O'Connell and Michael Ruse

Adaptation
Elisabeth A Lloyd

Stem Cells
Melinda Bonnie Fagan

A full series listing is available at www.cambridge.org/EPBY

Printed in the United States
by Baker & Taylor Publisher Services

Printed in the United States
by Baker & Taylor Publisher Services